AF522290

Crop Production: Theory, Techniques and Technology

Crop Production: Theory, Techniques and Technology

Prof. Sriniwas Puri

RANDOM PUBLICATIONS
NEW DELHI (INDIA)

Crop Production: Theory, Techniques and Technology

ISBN 978-93-5111-598-4

Published in 2015 in India by

RANDOM PUBLICATIONS

4376-A/4B, Gali Murari Lal, Ansari Road
New Delhi-110 002
Phone : +9111-43580356, 011-23289044, 011-43142548
e-mail: sales@randompublications.com,
info@randompublications.com, randomexports@gmail.com

Reprinted 2026

Type Setting by : Friends Media, Delhi-110089

Preface

Crop production depends on the successful implementation of the soil, water, and nutrient management technologies. Food production by the year 2020 needs to be increased by 50 per cent more than the present levels to satisfy the needs of around 8 billion people. Much of the increase would have to come from intensification of agricultural production.

Crop production is a complex business, requiring many skills (such as biology, agronomy, mechanics, and marketing) and covering a variety of operations throughout the year.

Finally then, one of the great principles of crop production is that there is much more to it than adaptation to the biophysical environment in which crops grow. Understanding the basics of plant growth is an important beginning, but successful cropping is a game of measurement and decision-making that aims to balance odds in favour of farmers. Technology and management skill are essential if farmers are to conquer the challenge of variable climatic and economic environments.

Crop improvement technologies improve the potentialities of higher yield and food quality in crops. New varieties and hybrids of various crops released by various crop improvement techniques are ultimately aimed to increase the yield and better quality.

But, these potentialities are realisable only when crops are managed and grown under conditions conducive to express their genetic potentialities. The techniques to release the potentialities of varieties and hybrids of various crops are termed as crop production techniques.

The book includes scientific principles as well as technological advances made in this area of the crop production.

I would like to thank my team for standing beside me throughout my career and writing this book. My special thanks go to "Random Publications" who have published the book.

– Prof. Sriniwas Puri

Contents

1

Modern Concepts and Advances Principles in Crop Production

CROP PRODUCTION

Crop production is a complex business, requiring many skills (such as biology, agronomy, mechanics, and marketing) and covering a variety of operations throughout the year. In this module, the practice of crop production will be described by discussing eight components in the crop production cycle:

Fig. Natural Resources Conservation Service

For each component, the operations and when they need to be carried out, the machinery or equipment farmers use, potential environmental concerns related to that component, and best management practices recommended to minimize environmental problems will be described.

BACKGROUND

Growing crops for food was one of the first priorities of the earliest settlers arriving in North America. With shipboard supplies depleted, and having little familiarity with the land and native vegetation, groups arriving from Europe were quickly forced to learn to produce crops to ensure their survival. The stories of Native Americans teaching the settlers to plant and fertilize a corn crop are part of this country's lore.

In the era of Thomas Jefferson (arguably the most illustrious farmer that this nation has produced), farmers made up about 90 per cent of the workforce. As late as 1900, almost 40 per cent of the labour force was engaged in producing crops and livestock for food, feed, and fibre. Now, with less than one per cent of our population claiming farming as a principal occupation, most U.S. citizens have little or no crop production experience. This chapter provides an overview of the principles and practices associated with production of the major crops grown in the United States.

MAJOR CROPS GROWN IN THE UNITED STATES

In round numbers, U.S. farmers produce about $100 billion worth of crops and about $100 billion worth of livestock each year. Production data from the year 2000 for major agricultural crops grown in this country are highlighted in the following table:

Major agricultural crops produced in the United States in 2000 (excluding root crops, citrus, vegetable, etc).

Crop	Harvested Area (million acres)	Cash Receipts from Sales ($ billion)
Corn (grain)	72.7	15.1
Soybeans	72.7	12.5
Hay	59.9	3.4
Wheat	53.0	5.5
Cotton	13.1	4.6
Sorghum (grain)	7.7	0.82
Rice	3.0	1.2

Corn: The United States is, by far, the largest producer of corn in the world. Corn is grown on over 400,000 U.S. farms. In 2000, the U.S. produced almost ten billion bushels of the world's total 23 billion bushel crop. Corn grown for grain accounts for almost one quarter of the harvested crop acres in this country. Corn grown for silage accounts for about two per cent of the total harvested cropland or about 6 million acres. The amount of land dedicated to corn silage production varies based on growing conditions. In years that produce weather unfavorable to high corn grain yields, corn can be "salvaged" by harvesting the entire plant as silage.

According to the National Corn Growers Association, about eighty per cent of all corn grown in the U.S. is consumed by domestic and overseas livestock, poultry, and fish production. The crop is fed as ground grain, silage, high-moisture, and high-oil corn. About 12 per cent of the U.S. corn crop ends up in foods that are either consumed directly (*e.g.* corn chips) or indirectly (*e.g.* high fructose corn syrup). It also has a wide array of industrial uses including ethanol, a popular oxygenate in cleaner burning auto fuels.

Soybeans: Approximately 2.8 billion bushels of soybeans were harvested from almost 73 million acres of cropland in the U.S. in 2000. This acreage is roughly equivalent to that of corn grown for grain. Over 350,000 farms in the United States produce soybeans, accounting for over 50 per cent of the world's soybean production and $6.66 billion in soybean and product exports in 2000. Soybeans represented 56 per cent of world oilseed production in 2000.

Soybeans are used to create a variety of products, the most basic of which are soybean oil, meal, and hulls. According to the United Soybean Board, soybean oil, used in both food manufacturing and frying and sautéing, represents approximately 79 per cent of all edible oil consumed in the United States. Soybean oil also makes its way into products ranging from anti-corrosion agents to Soy Diesel fuel to waterproof cement. Over 30 million tons of soybean meal are consumed as livestock feed in a year. Even the hulls are used as a component of cattle feed rations.

Hay: Hay production in the United States exceeds 150 million tons per year. Alfalfa is the primary hay crop grown in this country. U.S. hay is produced mainly for domestic consumption although there is a growing export market. According to the National Hay Association, the most common exports are timothy, some alfalfa, sudangrass, and bermudagrass hay. Hay can be packaged in bales or made into cubes or pellets. Hay crops also produce seeds that can be used for planting or as specialized grains.

Wheat: Over 240,000 farms in the United States produce wheat. The U.S. produces about 13 per cent of the world's wheat and supplies about 25 per cent of the world's wheat export market. About two-thirds of total U.S. wheat production comes from the Great Plains.

Wheat is classified by time of year planted, hardness, and colour (*e.g.* Hard Red Winter (HRW)). The characteristics of each class of wheat affect milling and baking when used in food products. Of the wheat consumed in the United States, over 70 per cent is used for food products, about 22 per cent is used for animal feed and residuals, and the remainder is used for seed.

Cotton: Fewer than 32,000 farms in the United States produce cotton. Cotton is grown from coast-to-coast, but in only 17 southern states. Farms in those states produce over 20 per cent of the world's cotton with annual exports of more than $3 billion. The nation's cotton farmers harvest about 17 million bales or 7.2 billion pounds of cotton each year.

Cotton is used in a number of consumer and industrial products and is also a feed and food ingredient. Over 60 per cent of the annual cotton crop goes into apparel, 28 per cent into home furnishings, and 8 per cent into industrial products each year. Cottonseed and cottonseed meal are used in feed for livestock, dairy cattle, and poultry. Cottonseed oil is also used for food products such as margarine and salad dressing.

Grain sorghum: In the United States, grain sorghum is used primarily as an animal feed, but is also used in food products and as an industrial feedstock. Industrial products that utilize sorghum include wallboard and biodegradable packaging materials. Worldwide, over half of the sorghum grown is for human consumption.

Some farmers grow sorghum as a hedge against drought. This water-efficient crop is more drought tolerant and requires fewer inputs than corn. Kansas, Texas, Nebraska, Oklahoma, and Missouri produce most of the grain sorghum grown in this country. The U.S. exports almost half of the sorghum it produces and controls 70 per cent to 80 per cent of world sorghum exports.

As much as 12 per cent of domestic sorghum production goes to produce ethanol and its various co-products. With demand for renewable fuel sources increasing, demand for co-products like sorghum-DDG (dry distillers grain) will increase as well due the sorghum's favorable nutrition profile.

Rice: Just over 9,000 farms produce rice in the United States. Those farms are concentrated in six states: Arkansas, California, Louisiana, Mississippi, Missouri, and Texas. U.S. rice production accounts for just over 1 per cent of the world's total, but this country is the second leading rice exporter with 18 per cent of the world market.

About 60 per cent of the rice consumed in the U.S. is for direct food use; another 20 per cent goes into processed foods, and most of the rest into beer.

CROP PRODUCTION SYSTEMS

Major agricultural crops produced in the United States in 2000.

Crop	Harvested Area (million acres)	Cash Receipts from Sales ($ billion)
Corn (grain)	72.7	15.1
Soybeans	72.7	12.5
Hay	59.9	3.4
Wheat	53.0	5.5
Cotton	13.1	4.6
Sorghum (grain)	7.7	0.82
Rice	3.0	1.2

Of the seven crops listed, six are annual crops that must be replanted each year (only hay crops would be left in place from year to year). The process of cultivating crops typically begins with tillage of the soil. Although tillage can serve a number of functions within a crop production system, the most fundamental function is to create conditions that will ensure good contact between seed and soil at the time of seed planting and the ready availability of water to the seed during germination. The degree to which the soil is disturbed by tillage prior to seed planting provides a means of categorizing crop production within a range of tillage systems. These systems range from *no-tillage* in which there is not soil disturbance in a field except during the process of planting a crop to *conventional tillage* in which multiple tillage operations can extend over many months and take place before, during, and after planting. Crop production systems that involve pre-plant tillage but maintain residues from a previous crop on the soil surface are referred to as *conservation tillage* practices.

For the major row crops produced in the United States, farmers use a range of production practices. Conventional tillage (also known as*intensive tillage*) usually involves a series of field operations that result in a residue-free soil surface at the time a crop is planted. Conventional tillage systems developed in this country to take advantages of the following benefits:

- Creation of a seedbed or root bed
- Control of weeds or the removal of unwanted crop plants
- Incorporation of plant residues into the soil profile
- Incorporation of fertilizers and/or soil-applied pesticides
- Establishment of specific soil surface configurations for planting, irrigating, drainage, and/or harvesting operations

The major disadvantage of conventional tillage is the susceptibility of "unprotected" soil to erosion by water or by wind. Tillage is also energy-intensive, requiring large inputs of machine work and numerous trips across a field during a single growing season. Conventional tillage was "standard operating procedure" in the era before effective chemical weed and pest control strategies were available to farmers.

Concerns about soil erosion led to the development of crop production strategies that retained crop residues on the soil surface. Conservation tillage requires more sophisticated implements that are capable of producing a seedbed while leaving a portion of surface residues undisturbed. Reduced tillage usually leaves 15 per cent to 30 per cent residue coverage on the soil surface. True conservation tillage is any tillage method that leaves at least 30 per cent residue coverage on the soil after a crop has been planted. It can be accomplished through no-till, strip-till, ridge-till, ormulch till practices.

Organic Farming

Organic farming is a small, but growing, segment of U.S. agriculture. USDA estimates the value of retail sales of organic foods at $6 billion in 1999 with about 12,200 organic farmers nationwide, most with small-scale operations. Organic farming encompasses both crop and animal production and is defined as "ecological production management system that promotes and enhances biodiversity, biological cycles and soil biological activity."'Organic' is a labeling term that denotes products produced under the authority of the U.S. Organic Foods Production Act. "The principal guidelines are to use materials and practices that enhance the ecological balance of natural systems. Organic agriculture practices do not ensure that products are completely free of residues; however, methods must be used to minimize contamination." Organic food handlers, processors and retailers must adhere to standards that maintain the integrity of organic agricultural products. This includes practices such as minimizing or eliminating the use of herbicides in crop production and antibiotics in animal production.

SOIL PREPARATION

Prior to planting, the soil needs to be prepared, usually by some form of tillage or chemical "burn-down" to kill the weeds in the seedbed that would crowd out the crop or compete with it for water and nutrients. Tillage methods can be divided into three major categories, depending on the amount of crop residue they leave on the surface. Residue slows the flow of run-off that can displace and carry away soil particles.

- Conventional tillage - Until the last decade or so the standard tillage practice for corn was use of the moldboard plow for primary tillage followed by several secondary tillages and mechanical cultivation after the crop was up. Now about two-thirds of row crops are planted without use of the moldboard plow, and mechanical cultivation is often limited to one, or no operations.
- Reduced tillage is usually done with a chisel plow and leaves 15 per cent to 30 per cent residue coverage on the soil.
- Conservation tillage leaves at least 30 per cent residue coverage on the soil. Conservation tillage methods include no-till, where no tillage is done at all and seeds are placed directly into the previous season's

crop residue; strip-till, in which only the narrow strip of land needed for the crop row is tilled; ridge till; and mulch till.

Herbicides are used in all these methods to kill weeds. In no-till systems, the herbicide is applied directly on last season's crop residue. In the other methods, some soil preparation takes place before the herbicide is applied. A common myth is that more herbicide is used with conservation tillage methods, but in fact farmers rely on herbicides for weed control under all tillage systems, and the amount used is more or less independent of tillage method.

SOIL PREPARATION OPERATIONS AND TIMING

Tillage can occur anytime between harvest of the previous year's crop and spring planting. In the eastern Corn Belt, most tillage is usually done between March and May for corn, and can be as late as early June for soybeans. In some cases, tillage is done in the fall, after harvest.

In southern states, planting can be considerably earlier or later because of their longer growing season. The optimum time for tillage (to prevent soil erosion) is just before planting. However, wet spring weather can often make it difficult to get equipment into the field as early as needed to optimize yield. Late planting can seriously reduce yields. For example, in the eastern corn belt, corn yields are reduced by 1 bu/acre for each day after May 1 that planting is delayed.

EQUIPMENT USED FOR SOIL PREPARATION

Tractor - a traction machine that provides mechanical, hydraulic, and/or electrical power to implements to perform a wide range of crop production and handling operations. Tractors are most often used to perform drawbar work (pulling equipment through the field) and PTO (power take-off) (power to rotate equipment components) work. Tractors can be equipped with rubber tires, rubber belts, or steel tracks.

Fig. Farm tractor and tillage implementation

A modern farm tractor is almost always equipped with a diesel engine and tractor size is measured by the amount of power that the tractor can produce at the PTO.

Tractor sizes range from those with less than 40 PTO horsepower to ones that produce more than 400 horsepower. The cost of a large modern tractor can be well over $200,000.

Fig. Rubber-tracked tractor and moldboard plow at work in the field

Plow - an implement used to perform primary tillage. A number of types of plows are in common use including the *moldboard plow*, the*chisel plow*, and the *disk plow*.

The moldboard plow has a large frame that is equipped with a series of "bottoms," each of which consists of a steel coulter to slice through residue followed closely by a steel share that cuts the soil and an attached moldboard that is used to raise and turn over the cut "slice" of soil.

Disk plows work in a similar manner to laterally displace and invert soil through the use of concave steel disk blades.

Chisel plows use curved shanks to penetrate and "stir" the soil without inverting a soil layer. Chisel plows cause less residue disturbance than moldboard plows and are often used in conservation tillage systems.

Fig. A close-up view of a disk harrow in the field

Disk Harrows (or Disk) - are implements that uses steel blades to slice through crop residues and soil. Disk blades are mounted in groups or gangs that rotate as they move forward through the soil.

Front gangs move soil towards the outside of the disk while rear gangs move soil back towards the center of the disk. A disk can be used for primary or secondary tillage.

Fig. A field shot of a tractor and disk harrow at work

Field Cultivator -an implement used to perform secondary tillage operations such as seedbed preparation and weed eradication. Field cultivators are equipped with steel shanks that are typically spring mounted to permit the shank to move within the soil and shatter clods.

Field cultivators are constructed similarly to chisel plows, but are more lightly built. Large chisel plows can exceed 50 feet in width in the field.

Fig. A tractor and row crop cultivator working in soybeans plated with a conservation tillage system

ENVIRONMENTAL CONCERNS RELATED TO SOIL PREPARATION: SOIL EROSION

The major environmental concern related to soil preparation is erosion. Soil erosion is a natural process that occurs when the actions of water and/or wind cause topsoil to be removed and carried elsewhere.

Soil erosion can be caused by either water or wind. In many agricultural areas, soil is eroding at a rate of several tons of soil per acre per year or higher. The map shows an estimate of total soil erosion on agricultural areas in 1992. This includes both cropland and set-aside land in the Conservation Reserve Programme. Forested and urban land is not included in the map.

The good news is that soil erosion in the U.S. is decreasing. From 1982-1997, soil erosion declined about 40 per cent in the U.S., due to government conservation programmes, technological advances, and extension education efforts.

Water erosion is caused by the erosive power of raindrops falling on the soil (particularly if the soil is not covered by vegetation or residue) or by surface run-off. Raindrops cause the less severe forms of erosion (know as sheet and interrill erosion). Severe erosion problems such as rill erosion, channel erosion, and gully erosion can result from concentrated overland flow of water.

Wind erosion is particularly a problem in windy areas when the soil is not protected by residue cover. Wind erosion in the United States is most widespread in the Great Plains states, as can be seen in the map at right. Wind erosion is a serious problem on cultivated organic soils, sandy coastal areas, alluvial soils along river bottoms, and other areas in the United States.

Impacts of soil erosion

Soil erosion has both on-farm impacts (reduction in yield and farm income) and off-farm impacts (contaminated water due to the sediment and associated contamination from nutrients and pesticides carried on the soil particle).

On-farm impacts due to the loss of soil and nutrients include:

- Lower fertility levels
- Development of rills and gullies in the field
- Poorer crop yields
- Less water infiltration into the soil
- More soil crusting
- More run-off in the spring and after storms

When fertile topsoil is lost, nutrients and organic matter needed by crops often are removed along with it. Erosion tends to remove the less dense soil constituents such as organic matter, clays, and silts, which are often the most fertile part of the soil. However, the loss in productivity caused by erosion has not been so evident in many parts of the U.S., since it has been compensated for over the years by improved crop varieties and increased fertilization.

Soils can tolerate a certain amount of erosion without adverse effects on soil quality or long-term productivity, because new soil is constantly formed to replace lost soil. This tolerable level is known as "T" and generally ranges from 3 to 5 tons per acre per year. Goals for reducing soil erosion often use the "T" value as a target, because erosion rates below T should maintain long-term productivity of the soil.

Off-farm impacts occur when the eroded soil is deposited elsewhere, along with nutrients, pesticides or pathogens that may be attached to the soil. The tolerable"T" value described above does not take into consideration the off-farm or downstream impacts. Soil eroded by water has effects such as:

- Eroded soil deposited in depressions and adjacent fields
- Decreased water quality downstream
- Decline of downstream aquatic ecosystems because of sedimentation and the addition of nutrients, pesticides, and bacteria associated with the soil
- Clogged drainage ditches and other costly problems

Off-farm impacts of wind erosion are due to the blowing soil, which can reduce seedling survival and growth (seed cover), increase the susceptibility of plants to certain types of stress, contribute to transmission of some plant pathogens, and reduce crop yields. Dust affects air quality, obscures visibility which can cause automobile accidents, clogs machinery, and deposits in road ditches, where it can impact water quality.

BEST MANAGEMENT PRACTICES TO REDUCE EROSION

Conservation tillage leaves at least 30 per cent residue cover on the ground. This simple, low-cost practice can have a huge impact on the amount of soil eroded. Because of energy savings and obvious improvements in soil quality that can result from conservation tillage, it has been widely adopted across the Midwest. In Indiana, for example, conservation tillage was used on 50 per cent of corn and 80 per cent of soybean acres in 2000, a dramatic improvement from 10 years earlier. There is still room for improvement, however. This map shows the per cent of U.S. crop land currently in conservation tillage. Percentages are generally higher for soybeans than for corn or other crops.

Contour farming and strip cropping is the practice of planting along the slope instead of up-and-down slopes, and planting strips of grass between row crops. Cover crops are crops such as rye that grow in late fall and provide soil cover during winter. By providing a cover to the soil, winter soil erosion from both air and water can be greatly reduced.

Grassed waterways protect soil against the erosive forces of concentrated run-off from sloping lands. By collecting and concentrating overland flow, waterways absorb the destructive energy that would otherwise cause channel erosion and gully formation. Terraces are structural practices that can reduce erosion by holding back the water and routing it along a channel at a lower velocity to where it can be safely discharged, usually into a grassed waterway.

Windbreaks are the best way to protect soil from wind erosion. They can be in the form of rows of shrubs or trees.

Windbreaks

Grass barriers can prevent wind erosion by slowing the wind.

"Living snow fences" prevent wind erosion by slowing the wind.

IRRIGATION

Irrigation makes agriculture possible in areas previously unsuitable for intensive crop production. Irrigation transports water to crops to increase yield, keep crops cool under excessive heat conditions and prevent freezing.

Less than 15 per cent of U.S. cropland is irrigated, although irrigation is essential for crop production in some of the most productive areas of the country. For instance, in Arizona, home to some the highest corn yields in the country (208 bushel per acre state average in 2001 compared to 152 for Illinois), much of the crop is under continuous irrigation from planting until harvest.

The need to irrigate is usually driven by the necessity to meet the water needs of the crop from year to year (some areas of the country simply receive too little rainfall during the growing season to support economical crop growth). In other situations, irrigation is viewed as insurance against occasional drought. In areas where rainfall is plentiful in most years, irrigation can bring benefits by reducing risk, meaning that a farmer is better able to control income fluctuation. Other benefits include:

- Improving crop quality (most noticeable for vegetable crops)
- Significantly increasing crop yields, particularly on sandy soils which have low moisture-holding capacities
- Increasing opportunities for double cropping (planting soybeans after wheat in the same year)
- Providing a means of liquid fertilizer application

In 1997 there were about 55 million irrigated crop acres in the U.S. Irrigation is concentrated in certain areas like central California, Nebraska and the Great Plains, and the lower Mississippi valley.

Although irrigation has always been most common in the West, U.S. irrigated acreage in the East has also grown from 11 per cent of acres in 1969 to 22 per cent of acres in 1997.

Irrigation water is obtained from either ground water or surface water. Wells drilled on the farm are a common source of water in many areas, and are usually the only source used in the Great Plains. Offsite sources such as rivers, pipelines, canals operated by irrigation districts and private water companies, are also used, mainly in western states. The percentage of water source used for irrigation varies across the U.S.

OPERATIONS AND TIMING

Irrigation water is applied throughout the growing season to meet crop needs. Moisture needs depend on the type of crop and its stage of development. In the Eastern Corn Belt, for example, it takes 20-22 inches to produce an optimal corn crop, 18-20 inches for a soybean crop, 12-13 inches for small grain, and 24-26 inches for alfalfa.

Irrigation can reduce crop stress if rainfall does not provide this amount of moisture during the growing season.

It is not only total moisture, but also the timing of moisture application (or rain) that is necessary for optimum crop yields. Crops have critical periods during the growing season when soil moisture must be maintained to ensure optimal yields.

For corn, the most critical period is from just before tasseling through silking. For small grain, it is from boot to heading stage, for alfalfa, the start of flowering and after cutting, and for pasture, after grazing.

EQUIPMENT USED

There are four primary types of irrigation:

Surface Irrigation - With surface irrigation, water flows directly over the surface of the soil. The entire surface can be flooded (most often used for crops that are sown, drilled, or seeded) or the water can be applied through furrows between the rows (for row crops).

Fig. Center Pivot Irrigation

Sprinkler Irrigation - With sprinkler irrigation, water is sprayed through the air from pressurized nozzles, and falls like rain on the crop.

Variable-Flow Irrigation - Variable-flow irrigation sprinkler head improves the precision of water and farm chemical applications.

Trickle or Drip Irrigation - Trickle or drip irrigation supplies water directly onto or below the soil surface through "emitters' that control water flow.

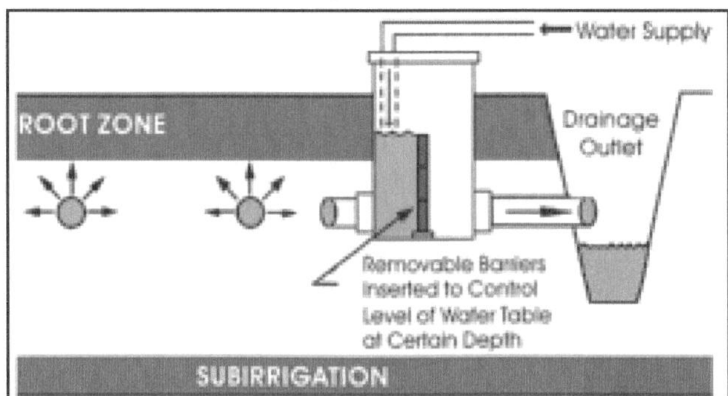

Fig. Modified from graphic by Leslie Zucker, Ohio State University Extension

Subirrigation - With subirrigation, the water table is artificially raised either through blocking ditches or by supplying water through the perforated pipes also used for subsurface drainage.

Irrigation types can be further distinguished by whether the equipment is permanently installed in one place (stationary system) or whether it is used until the necessary amount of water is applied, then moved to a different area (traveling system). Stationary systems such as permanent spray installations or trickle systems require less labour, but usually cost much more to install. Traveling systems such as center pivot sprinkler irrigation, linear-move, or cable-tow require more labour but less capital expense.

POTENTIAL ENVIRONMENTAL CONCERNS

Environmental concerns related to irrigation include depletion of the water source (falling water tables or reduced water levels in streams or reservoirs), soil erosion due to over-application, run-off and leaching of chemicals, and salinization of the soil (salt-buildup) and minerals and nutrients in the irrigation return flow that drains from the irrigated area.

BEST MANAGEMENT PRACTICES

1. Minimize water use. Apply only enough water to meet crop needs. This can be determined through regular soil moisture monitoring or through a "checkbook" system to monitor water applied and crop needs.
2. Irrigation efficiency. Use efficient irrigation systems such as drip irrigation to minimize evaporation.
3. Apply at rate the soil can absorb. Run-off due to excess irrigation can cause soil erosion.
4. Uniform Irrigation. Make sure water is applied uniformly. This makes the water more efficient, and reduces the chance of run-off and leaching in certain areas where water may be overapplied.
5. Provide good drainage. Salinization in areas of low rainfall can be minimized by providing good drainage along with the irrigation, to leach salts down through the soil profile.

DRAINAGE

Fig. Impacts of poor drainage — stunted, yellow plants

The purpose of agricultural drainage is to remove excess water from the soil in order to enhance crop production. In some soils, the natural drainage processes are sufficient for growth and production of agricultural crops, but in many other soils, artificial drainage is needed for efficient agricultural production.

Surface drainage is the removal of water that collects on the land surface. Many fields have low spots or depressions where water ponds. Surface drainage techniques such as land leveling, constructing surface inlets to subsurface drains, and the construction of shallow ditches or waterways can allow the water to leave the field rather than causing prolonged wet areas.

Fig. Poorly drained area in crop field will damage yields

Subsurface drainage removes excess water from the soil profile, usually through a network of perforated tubes installed 2 to 4 feet below the soil surface. These tubes are commonly called "tiles" because they were originally made from short lengths of clay pipes known as tiles.

Water would seep into the small spaces between the tiles and drains away.

Fig. Drain tile outlet to a drainage ditch

The most common type of "tile" is corrugated plastic tubing with small perforations to allow water entry. When the water table in the soil is higher than the tile, water flows into the tubing, either through holes in the plastic tube or through the small cracks between adjacent clay tiles. This lowers the water table to the depth of the tile over the course of several days. Drain tiles allow excess water to leave the field, but once the water table has been lowered to the elevation of the tiles, no more water flows through the tiles. In most years, drain tiles are not flowing between June and October.

OPERATIONS AND TIMING

On average, about two-thirds of annual precipitation is used by crops in the eastern Corn Belt. The rest falls at a time when it does not meet crop needs.

Monthly precipitation remains fairly constant throughout the year, while evapotranspiration (a combination of evaporation from soil and transpiration from the crop), is much higher from June to September. From January to May, and from October to December, precipitation is greater than evapotranspiration, creating a water surplus.

The surplus results in excess water in the crop root zone and the need for drainage. Drainage is primarily a concern in the periods prior to the growing season (January to April) so that crops can be planted at the optimum time.

EQUIPMENT USED

Fig. Trenching machine used by drainage contractors to install subsurface drainage tile

Fig. Drainage plow being pulled by farm tractor installing a tileline

Designing and installing a drainage system is a complex process. Every field is unique and usually requires an individual design. Drainage depends on topography, crops that will be grown on the field, and soil type. Every soil type has different properties that affect its drainage. Agronomists and engineers have developed recommendations for drainage depth and spacing in specific soil types based on years of experience and knowledge of soil properties. Drainage contractors use these recommendations to design drainage systems that economically and effectively drain a particular field.

Drainage plows that can be pulled by farm tractors are becoming more popular. But most farmers hire contractors to design and install their tile drainage systems because of the knowledge, skills, and experience needed to install a successful system.

POTENTIAL ENVIRONMENTAL CONCERNS

The major concerns related to drainage are:

- Loss of wetlands, and
- Increased loss of nitrate through tile drains.

Wetlands

Much of the Midwest landscape consisted of wetlands before large-scale drainage began in the 19th century. Although enormous public health and economic benefits have resulted from the draining of these wetlands over the last 150 years, there have also been negative impacts on the environment. Wetlands have an important hydrologic function in regulating water flow and maintaining water quality, as well as providing habitat for water-based wildlife. Recognition of their value has changed the way our society thinks about and protects wetlands.

Drainage improvements today are rarely for the purpose of converting existing wetlands to agricultural production. Improved drainage is usually aimed at making existing agricultural land more productive.

Some fields have drain tiles that were installed 100 or more years ago, and are broken or plugged. In many fields, only a few of the wettest spots were originally drained, while the entire field would benefit from improved drainage. More tiles are often added to improve drainage efficiency, with the goal of increasing production.

WATER QUALITY

Fig. Poor Drainage

Drainage has both positive and negative effects on water quality.

In general, less surface run-off, erosion, and phosphorus is lost from land that has good subsurface drainage than from land without drainage improvements or with only surface drainage.

Nitrate loss can be quite high from drained land. Because nitrate is very soluble, it flows easily through the soil and into tile lines. Nitrate flow from subsurface drains is one of the main sources of nitrate in streams and rivers in the Midwest.

Concern about hypoxia, or low oxygen levels, in the Gulf of Mexico has increased concern about this nitrate source. Concentrations of nitrate in tile drains are usually quite high (10-40 mg/l).

Pesticides can also flow into subsurface drains, but usually only in very low concentrations. Pesticides move more easily in flow over the soil surface than through the soil, so the highest concentrations of pesticides in tiles are often in fields that have surface inlets into the drains. In fact, subsurface drainage may actually reduce pesticide loss to rivers and streams because it reduces surface run-off.

BEST MANAGEMENT PRACTICES

Traditionally, the goals of drainage were to:

1. Maximize crop yield and
2. Minimize costs of drainage installation.

Reducing water quality effects of drainage is becoming a third objective in drainage design.

Nitrate loss is the biggest water quality concern related to tile drainage. Several new technologies can reduce nitrate loss. Controlled drainage keeps the water table high during the off-season when crops are not growing. The high water table increases the rate of denitrification (a process that converts nitrate to harmless nitrogen gas (N_2) as soon as the saturated soil warms up in the spring) and reduces nitrate loss to the environment.

Controlled drainage can be combined with subirrigation to improve yields while protecting water quality. Subirrigation is irrigation back through the subsurface drain tiles.

Subirrigation may be economical when fields are relatively level and need to be drained anyway, since additional infrastructure consists mainly of increased numbers of tiles the pumping system.

One system being developed in Ohio combines a wetland for water treatment and a pond serving as a reservoir for subirrigation with a drainage system. This system has been shown to increase yields and reduce water quality impacts of drainage, although it is costly.

HARVEST

Harvesting crops involves getting the crop out of the field and transported to market. Most crops are harvested in the fall, except for hay which is cut several times over the course of the summer.

Field crops are harvested by machine, while small fruits and other food crops are typically harvested by hand, although in certain cases, they may be harvested by machine.

OPERATIONS AND TIMING

Crop	Typical harvest period (in eastern corn belt)
Corn - grain	October 7 to November 3
Corn silage	September 1 to October 15
Soybeans	October 1 to October 20
Wheat (spring)	August 14 to September 1
Wheat (winter)	June 15 to July 15
Hay	Usually 3 cuttings from May 15 to Sept. 30

EQUIPMENT USED

Tractors - are traction machines that provide mechanical, hydraulic, and/or electrical power to implements to perform a wide range of crop production and handling operations. Tractors are most often used to perform drawbar work and PTO (power take-off) work.

Fig. Farm tractor and tillage implement

Tractors can be equipped with rubber tires, rubber belts, or steel tracks. A modern farm tractor is almost always equipped with a diesel engine and tractor size is measured by the amount of power that the tractor can produce at the PTO. Tractor sizes range for those with less than 40 PTO horsepower to ones that produce more than 400 horsepower. The cost of a large tractor can exceed $200,000.

Fig. A self-propelled forage harvester loading chopped corn into a truck on-the-go.

Forage Harvesters - are tractor-drawn implements or self-propelled machines that are used to gather, chop, and discharge forage crops as it moves through the field.

Fig. Tractor and pull-type combine operating in a small grain crop

The crops are typically harvested at a very high moisture content to permit ensiling (preservation through anaerobic fermentation). Forage harvesters require a great deal of power to perform the required functions. The largest self-propelled forage harvesters currently available have diesel engines that produce in excess of 600 horsepower. Combines - are farm machines used to harvest grain and seed crops. The major functions performed by a combine include cutting and/or gathering, feeding, threshing, separating, cleaning, and grain handling operations on-the-go in the field. The vast majority are self-propelled, receiving power to perform all of the previously listed operations and traction from a diesel engine.

Fig. Self-propelled combine equipped with a grain header

Fig. Self-propelled combine equipped with a corn head preparing to harvest corn

Fig. Cotton harvesters at work in the field

The combine is often the most expensive farm machine used in grain or row crop production with list prices for the largest models exceeding $350,000. Some machines can harvest a thirty-foot swath of crop in a single pass through the field. Cotton Harvesters - are self-propelled machines specifically designed to pick (or strip), accumulate, and off-load large quantities of cotton in the field.

Fig. A large-capacity grain cart unloading into a waiting truck

Grain Carts - are tractor-drawn implements used to shuttle grain from combines to hauling vehicles or to grain receiving facilities. Grain carts are usually equipped with "high-flotation" tires or rubber tracks to attempt to minimize soil compaction in the field. The capacity of such carts can exceed 1,000 bushels (equivalent to 56,000 lb of shelled corn or 60,000 lb of soybeans).

Fig. Large round baler discharging a newly-formed bale of hay

Balers - are implements used for packaging hay, or straw to permit mechanized handling and transport.

Fig. Large rectangular baler discharging a newly-formed bale of straw

The two most common bale formats are large round bales (> 4 feet in diameter) and large rectangular bales (up to 8 feet in length). Large bales can weigh more than 2,000 pounds.

STORAGE OF FERTILIZER AND CHEMICALS

FERTILIZER AND PESTICIDE STORAGE

Summary of an Ideal Fertilizer and Pesticide Storage Facility

Properly designed facilities promote storage, handling and disposal practices that enhance worker safety and minimize the risk of point source contamination. An ideal facility incorporates safety features in all aspects of its design and provides:

- Secure storage of fertilizers away from pesticides.
- Secondary containment of day to day spills resulting from normal mixing/loading operations.
- Secondary containment of large, accidental spills or leaks (separate secondary containment for pesticides and fertilizers)
- Facilities for collecting, storing and recycling excess spray solutions and rinsates.
- A dry, secure, well managed area for storing empty containers and other waste prior to proper disposal.
- Office facilities for effective management and communications.
- Orderly, accessible storage for personal protection equipment (PPE) and emergency supplies.
- Worker convenience facilities: first-aid and training areas, restrooms, shower(s), laundry.

BUILDING A NEW STORAGE FACILITY

Building a new storage facility just for fertilizer storage may be expensive, but generally it will be safer than trying to modify areas meant for other purposes.

When selecting a site on which to build the facility, two basic criteria should be considered:

1. Human safety
2. Environmental safety

When designing and building a new fertilizer or pesticide storage facility, keep in mind a few simple principles of safe storage:

1. Check with your state Cooperative Extension Service or department of agriculture for design recommendations and requirements of a fertilizer or pesticide storage facility.
2. Check the local and state building codes and local zoning requirements before constructing the facility.

3. Have the site checked for background levels for potential soil and water contaminants.
4. Draw a facility site plan on which you locate and sketch important structures and activity areas using accurate dimensions and distances.
5. Locate the building downslope and away from your well. Separation from the well should be greater if the site has sandy soils or fractured bedrock near the soil surface. If the site must be upslope from the well, be sure to take precautions to have any spill contained to prevent it from moving into the water that supplies your well.
6. Surface water should drain to a retention area in the event of a fire.
7. The mixing and loading area should be close to your storage facility, to minimize the distance that chemicals are carried.
8. Maintain safe separation distances from your fertilizer or pesticide storage facility to other structures and activity areas.
9. The building foundation should be well drained and high above the water table. The finished grade should be 3 inches below the floor and sloped away from the building to prevent frost heave. The subsoil should have a low permeability.
10. Provide pallets to keep large drums or bags off the floor in order to keep them dry. Shelves for smaller containers should have a lip to keep the containers from sliding off easily. Steel shelves are easier to clean than wood if a spill occurs.
11. If you plan to store large tanks, provide a containment area capable of confining 110-125 per cent of the volume of the largest container (check state regulations).
12. The building may need to be insulated and heated if fertilizer is to be stored over the winter. It may need to be air conditioned in the summer, depending on the type of fertilizers or pesticides stored.
13. Keep the building locked and clearly labeled as a fertilizer or pesticide storage area. Provide exterior illumination of the warning signs and of the building to identify it as a fertilizer or pesticide storage facility. Preventing unauthorized use of fertilizers or pesticides reduces the chance of accidental spills or theft. Labels on the windows and doors of the building give firefighters information about fertilizers and other products present during an emergency response to a fire or a spill. It is a good idea to keep a separate list of the chemicals and amounts stored. If a fire should occur, consider where the water used to fight the fire will go and where it might collect. For example, a curb around the floor can help confine contaminated water.
14. Provide adequate road access for deliveries and use, and in making the storage area secure, also make it accessible, to allow getting fertilizers and other chemicals out in a hurry.

MODIFYING AN EXISTING FACILITY

Some of the principles for safe fertilizer storage:

1. Provide pallets to keep large drums or bags off the floor. Shelves for smaller containers should have a lip to keep the containers from sliding off easily. Steel shelves are easier to clean than wood if a spill occurs.
2. If you plan to store large bulk tanks, provide a containment area large enough to confine 125 per cent of the contents of the largest bulk container.
3. Keep the building locked and clearly labeled as a fertilizer storage area. Preventing unauthorized use of fertilizers reduces the chance of accidental spills or theft. Labels on the windows and doors of the building give firefighters information about fertilizers and other products present during an emergency response to a fire or a spill. It is a good idea to keep a separate list of the chemicals and amounts stored. If a fire should occur, consider where the water used to fight the fire will go and where it might collect. For example, a curb around the floor can help confine contaminated water
4. Provide adequate road access for deliveries and use, and in making the storage area secure, also make it accessible, to allow getting fertilizers and other chemicals out in a hurry.
5. Never store fertilizers inside a wellhouse or a facility containing an abandoned well.

If you decide to improve your current storage building, applying the above principles can be expensive. However, compared to the cost of a major accident, fine or lawsuit, storage improvements can be a bargain! If that is not practical, consider how you can protect the fertilizers that you keep on hand.

Sound containers are your first line of defence against a spill or leak. If a container is accidentally ripped open or knocked off a shelf, the spill should be confined to the immediate area and promptly cleaned up. The building should have a solid floor and, for liquid fertilizers, a curb. The containment volume should be large enough to hold the contents of the largest full container.

Ideally, the fertilizer storage area should be separate from other activities. If the building must also serve as a machine shed or as livestock housing, you may find it difficult to meet all the requirements for safe storage.

PRINCIPLES OF CROP PRODUCTION

Green plants occupy a special place among all living things on earth, the biosphere, because, uniquely, through their greenness, they connect earth to an external source of energy, the sun. They 'pack' this energy into sugars or carbohydrates (CHOs) that become the basis of all other living things, except for some chemoautotrophic organisms, principally bacteria. This puts them at first position on the food chain that includes all other organisms: humans, carnivorous and omnivorous creatures, and decomposers.

The carbon-based organic compounds that plants accumulate in vegetative structures and storage organs are, directly or indirectly, food for the vast majority of organisms in the biosphere. Plants alone are independent of other organisms by fixing carbon (C) from atmospheric carbon dioxide (CO_2) and producing oxygen (O_2) as well as carbohydrates (CHOs). They are dependent only on mineral nutrients and water they absorb from soil for the internal chemistry of their metabolism.

The unique process by which plants fix solar energy that then moves through the entire food chain until the C is returned as CO_2 to the atmosphere is called photosynthesis Humans are a part of the food chain and while our ancient ancestors collected plants and hunted animals, we now, with technological development and a large and growing population, secure our food supply and that of our domestic animals by growing plants specifically for those purposes. This essay is written to introduce this important topic of the role of crops in food supply for human existence and survival. While it is written for an agricultural audience, the general principles are relevant to all. This is true if only to counter a view, held by many urban dwellers, now comprising 50 per cent of world population, that food originates in supermarkets.

The essay is presented in two major sections, using a structure recommended by my Australian colleague David Smith as in his book "Rain and Shine". First is "How plants grow". David insists we keep the basics in mind at all times using understanding of How Plants Grow as a critical backdrop to farming, which is the second part of the essay, "How we Grow Plants". Backing these two sections is a vast body of scientific work that cannot be included here but is available to interested readers in books like "Crop Ecology: Productivity and Management in Agricultural Systems ".

HOW PLANTS GROW

Seed germination

Small though they often are, seeds are fascinating entities. As well as containing the 'germ' – the life package of genes/chromosomes that define the plant – they contain precursor chemicals that set growth and cell differentiation in train, and a small supply of the CHOs mentioned above. Thus, when seeds imbibe water, chemical and osmotic forces begin to operate so that embryonic cells divide and tissues of root and shoot begin to grow, recognizable in a day or two. The process is germination and seedlings can continue growth until the chemical reserves provided to them by mother plants are exhausted. The beginnings of a sense of geo-positioning in the world provided by a seed's genetic code are almost immediately obvious. First, root systems grow downwards and then shoot systems, in some species carrying the seed, grow upwards, pushing through any soil cover or litter into the atmosphere.

For individual seedlings, there are two critical steps that must be achieved before reserves are exhausted. First, emergence from the soil and expansion of seedling leaves and secondly, penetration of the developing root system into soil where it can absorb water and essential mineral nutrients. Then occurs the greatest miracle after life itself, seedling leaves turn green in the presence of sunlight by formation of chlorophyll and connect the plant to the power of the Sun. The plant is then an established properly functioning independent organism that can accumulate C-based compounds and the plant body can grow. Life on earth can now go on! These processes should be constantly on show to all children in primary school classrooms.

Continuing growth, increase in size and leaf area, requires a continuous supply of CHOs from photosynthesis in new expanding leaves and water and nutrients absorbed from soil by roots. But there is much more to growth than this, however. Plants grow by forming new organs, *viz.* roots, stems, seeds, and other storage organs. These processes are controlled by environmental responses that, like geotropism, come embedded in the genetic code. Thus, most plants will only flower and produce seed if conditions of temperature and day length are appropriate, and these conditions vary widely from species to species. These responses are adaptations that have evolved to ensure that, once established, development will normally lead to flowering, pollination, and seed production, thus maintaining the existence of the species. Survival of successful plants drives evolutionary adaptation. Most plants require pollination from another member of the species so the new seedling is 'after its kind' but not identical to either parent – therein lays scope for evolution – and human controlled plant breeding also.

DEVELOPMENT AND FLOWERING

Some plants are perennial and live for a number of years, in the case of some trees, for thousands. Many are annual and germinate when conditions are suitable for growth, have a herbage producing stage, then flower and set seed, usually avoiding the season when growing is difficult – too dry, too hot, or too cold. Thus flowering is a hugely important stage in plant development.

It follows changes in activity of some buds (meristems) found at ends of shoots, and in axils of leaves and branches, to form flowers rather than more leaves. These changes are mediated by chemical signals sent to these meristems, in response to temperature or changes in day length, or a combination of the two. For example, wheat and canola respond to increasing day length, so are long-day plants that flower in Spring, though some cultivars require a chilling period (low temperature) before they can respond to increasing day length.

So we have "winter" and "spring" types, suited to different sowing times depending upon the temperature regime. This terminology can be confusing, however. Winter types are always sown in Autumn, but in places where Winter is not severe, spring types can be sown also: where Winter is too severe for plant survival, however, only spring types sown in Spring can be successful.

The location of flower buds on plants is also important because it has an effect on growth strategies and thence seed production. Wheat and canola flower on terminal shoots – we call it 'determinate', field peas flower in axils – we call that 'indeterminate'. Each flowering trait has its own advantages and disadvantages.

For example, determinate crops are vulnerable to unusual, isolated, periods of stress during flowering. In wheat, for example, a single stress from frost or heat can abort most flowers, so that grain yield is at best very small. In contrast, prolonged flowering of indeterminate crops means that such short stress periods of frost or heat or water shortage have less effect on seed yield. To complicate the matter further, with indeterminate plants there is significant competition for assimilates (CHOs) between growth of seed and further vegetative parts. In determinate plants, however, once seed is set the bulk of assimilate goes to seed growth. Thus, where plants are growing under resource-limiting conditions determinate plants that make a sharp transition in growth pattern at the best time – *i.e.* the "optimum switchover" from vegetative to reproductive growth -will yield better. By contrast in indeterminate plants, vegetative material increases as yield increases and partition of growth between vegetative and reproductive parts is not optimal for high yield.

Photosynthesis

Photosynthesis is the process of incorporating CO_2 into CHOs in chloroplasts of green leaves. Though the beginning and end points are the same, green plants, all of which have evolved from simple algae, have one of three systems of photosynthesis with different combinations of chemical steps and associated plant anatomy. The most common is the original algal C_3 system (so named to recognize the chemical structure of the first product of photosynthesis), then two evolved variants called C_4 or CAM, respectively. Crop plants are mainly C_3 species.

All leaves have stomates (pores) through which CO_2 diffuses when photosynthesizing leaves reduce internal CO_2 below atmospheric concentration. In C_3 plants, arrangement of internal cells means that movement of CO_2 into chloroplasts, where photosynthesis occurs, is slow. In C_4 plants, by contrast, CO_2 is captured by an enzyme on entry and transported as C_4 acids to chloroplasts in specialized cells that surround vascular traces. There, CO_2 is released in high concentration and enters C_3 photosynthesis. The result is that C_4 plants have higher rates of photosynthesis than C_3 plants. This system also works better than C_3 at high temperature so, unsurprisingly, most C_4 crops are found naturally in tropical and subtropical climates. For example wheat, canola and field pea are C_3 plants while corn and green foxtail are C_4 plants.

The third system, crassulacean acid metabolism, CAM for short, is named after a family of succulent plants, the Crassulaceae, in which it predominates. This adaptation from C_3 is one of timing. Stomates are closed during day time but open at night when CO_2 enters and is concentrated in C_4 acids in large succulent cells. During daytime, CO_2 is released and enters C_3 photosynthesis. Photosynthetic capacity is limited by capacity for acid storage but some CAM plants also open stomates during the day and that increases photosynthesis. Pineapple is the most important CAM plant grown as a crop.

Water relations

As we know, plants require a continuous water supply. Although a small amount is used in chemical transformations, most is lost by evaporation from wet cells inside leaves and then by diffusion through stomates to the dry atmosphere beyond, in a process known as transpiration. Stomates must open to allow CO_2 entry to support photosynthesis so transpiration is an inescapable result of plant morphology and metabolism.

Plant growth is a continuing exchange with the atmosphere of water for CO_2 while plants seek to maintain water supply and turgor by extracting water from soil. The ratio of water loss:biomass gain is considerable and is in the range 800–1000:1. Since plants colonized land, evolution has presumably been seeking improvement in these processes, with limited but valuable results, such as in selection of C_4 and CAM types.

In CAM plants, the fact that stomates are mostly closed in daytime greatly reduces water loss so that the ratio of water loss to CO_2 gain over a daily period is much less than in C_3 and also in C_4 systems. Soils vary greatly in water holding capacity and plants differ significantly in rooting habit – depth, type *e.g.* taproots and fibrous roots -and ability to withdraw water from soil. They also vary in ability to control loss of water from leaves, owing to differences in size, shape and epidermal wax, and frequency and behaviour of stomates. Control of water loss is always made at the expense of CO_2 uptake for growth. For crops, flow of water from soil to atmosphere through plants is accompanied by direct evaporation of water from soil, particularly when the surface is wet and unprotected by foliage or mulch. This evaporation, which may consume as much of 50 per cent of water loss in annual crops, together with transpiration through plants, is summed in the term evapotranspiration. Reduction of soil evaporation saves water for transpiration so more plant growth can be supported.

Nutrition

Plant growth is a complex of chemical reactions that requires a continuing supply of essential mineral nutrients absorbed from soil by root systems. These nutrients are required in differing amounts, depending upon their roles in plant metabolism, so it is usual to distinguish between those required in large amounts, the macronutrients (N, P, K, S, Ca, Mg) from others in smaller amounts, the micronutrients (Si, Fe, Zn, Cu, Mn, Mo, B, Ni, Co, Cl). Macronutrients occupy central roles in growth, N in photosynthesis and protein metabolism, P in energy transfers, and K in osmotic relations of cells. Micronutrients are involved in a wide range of metabolic reactions. Fe, together with Mg, are critical elements in the structure and operation of chlorophyll and hence in photosynthesis. Cu and Mn are important for N-fixing bacteria that grow symbiotically with legumes.

Many people do not recognize that plant growth is essentially an extractive process taking minerals from the soil, the exact amount varying from plant to plant, amount of growth and stage of maturity. These minerals are in the soil because of break down of the parent rock, or through deposition by dust and/or water. Each plant takes some, and under continued cropping, extraction soon exceeds any new supply.

In nature, without human intervention all nutrients are returned to the system but in agriculture different amounts are extracted according to the composition of the harvested yield and its export. One interesting recent change, increasingly common in some countries for energy supply, has been harvesting of cereal straw as well as grain, leading to an additional large extraction of nutrients, notably of K.

Another interesting aspect is that though nutrients are ideally available in balanced amounts required by growing plants, there can be absorption and storage for later use during the growth cycle. Thus, in annual crop plants, N for construction of grain protein, more important in legumes and oilseeds that in cereals, is mobilized from chlorophyll in senescing leaves. Thus breeding for high yield in crop plants must keep in mind this evolutionary adaptation: the more rapidly N is extracted from leaves for grain growth, the less active leaf area remains for photosynthesis and assimilate production.

Nitrogen is a very special nutrient and there has been much emphasis on it in crop research and in programmes like the Green Revolution. Not only is N required in great amounts in plants but it is important through its relationship with the organic material of soil and chemical transformations by microbial activity there – and, ultimately with any sequestration of carbon in soils. The N content of decaying plant material is transformed to available, soluble, mineral form nitrate (NO_3^-) – but this is vulnerable to loss by leaching to depth as well as gaseous loss, mostly as N_2 to the atmosphere.

The major source of N in natural systems is from legumes that fix N_2 in symbiotic relationships with bacteria, mostly in root nodules. In agriculture this source is important but is complemented by addition of N fertilizer, some mined, but mostly extracted from the atmosphere using natural gas using the Haber-Bosch industrial process.

HOW WE GROW PLANTS

This leads us to a fairly brief discussion of human intervention in the above processes.

The change from hunter/gatherers to farming, even only as subsistence, was a logical one. Imagine a woman gathering food a long way from her cave sees a plant with a lot of fruit, picks some and it tastes good -better than any others she has tasted. Being a thinking person (that's what defines a human) she keeps the seed and uses her digging stick to bury it in soil nearer the cave, realizing this will mean a shorter walk to pick fruit in a year or two. The plant grows, and she is pleased. She protects her garden plants and to maintain their productivity removes others that crowd them. She and her family have noticed that faeces and urine are beneficial for plant growth, so they deposit them on the earth near plants. Here we see the beginnings of plant selection, soil cultivation, plant propagation, land clearing, and using manures, all in the cause of better production by plants and human convenience, "The beginning of agriculture".

In this way, subsistence farmers intervene in nature to grow more of the plants they like to eat or need for other uses. In the developed world, farmers grow plants that meet their economic requirements by reasons of price for product or cost of production. In this they compete for markets with other farmers, often in distant places. It is a fallacy to think that farmers simply grow the crops that are best suited to their biophysical environment. In fact, they select from plants that are sufficiently adapted to the environment (temperature, rainfall, evaporative demand) and manage them to avoid stresses that would seriously reduce yield. They do this by growing crops in sequences to make best use of biological interactions between plants of different types while also applying agrochemicals, when needed, to minimize risk of failure from low fertility and invasion by weeds, pests and diseases.

Cropping strategies

On individual farms, the sequences of crops that are grown define the cropping strategy chosen to meet economic and environmental objectives. Common features of the Prairie Zone now are sequences of cereal crops such as wheat and barley, with oil seeds such as canola and sunflower, and the legumes, field pea, lentil, and chickpea. Growing each particular crop means something different in terms of nutrient requirements, response to stresses, competitive ability with weeds, and susceptibility to herbicides applied and to their residues. Cereals are less extractive of nutrients than oilseeds or legumes and legumes fix nitrogen reducing the demand for N fertilizer. Stubble protects soil from erosion and evaporation, canola stubble inhibits root diseases of wheat, and so on.

The cropping strategy was not always like this. As in other parts of the world, prairie cropping systems are continuously evolving. The first settlers commenced with continuous cereals because that was the technology settlers brought with them.

As yields declined they introduced a biennial cropping sequence with an alternate (summer) fallow. Yields increased at first but gradually fell as fertility was extracted, and there was serious soil loss by wind and water erosion during sequences of years of low rainfall. New farming techniques and machinery enabled mulch farming to protect summer fallows, but they too have largely been replaced.

Research has defined the more important determinants of yield: control of weed burden and manipulation of the N cycle to optimise availability. In due course both were seen to be aspects of responses to fallowing (along with some soil moisture carryover), and it was realised the effects could be better and more economically achieved by using herbicides and N fertilizer or legume crops. Fallowing is now restricted to low rainfall areas where its contribution to water storage is aided by herbicides rather than tillage to control weeds.

Now with continuous cropping using reduced-tillage, even no-till techniques (conservation farming), soil organic matter and structure are being restored, and costs of land management and crop establishment greatly reduced. Machinery is larger and more precise so more of the operations can be completed with near optimal timing. Standing stubble protects soil from erosion and retains snow until Spring when crops are sown. Last year's stubble means seedlings from precision-sown seed are protected during establishment as they emerge between rows of stubble.

Fertilizers are applied as required and placed carefully relative to seed in appropriate quantities and formulations for most effective response. Changes to the systems continue as farmers adopt technology to adapt to changing environmental and economic conditions. The origins of change are often found in tactical variations in management as farmers seek to provide the best conditions for their crops allowing for year-to-year variability in weather and markets.

Tactical management

Measurement and technology and seasonal climate forecasts provide opportunities to better match crop performance with seasonal environmental conditions. This can either take advantage of high yield potential in some years or reduce costs when low yields are inevitable. Note we refer to cropping sequences and not "crop rotations" because that previously common term usually signified a specified, unbreakable sequence of crops. Tactical variation can start with adjustment to cropping sequence at sowing. Early or late start of season provides opportunity for change of cultivar or crop. Low content of water in the soil at sowing signals likelihood of low yield and inputs can be varied accordingly.

Low nitrogen at sowing signals need for fertilizer appropriate to water supply. In-crop measurements of soil water and crop N content, or related variables of leaf colour and tiller number, can provide the basis for late fertilizer application, or equally abandonment and diversion of crops to hay or silage. Herbicide applications can be adjusted to weed counts and specific weed populations that emerge with crops. Weather and short-term climate forecasts can be combined with crop models to predict outcomes as a basis for overall management. Farmers need to consider many issues to make the decisions that determine the viability of their cropping enterprises.

Finally then, one of the great principles of crop production is that there is much more to it than adaptation to the biophysical environment in which crops grow. Understanding the basics of plant growth is an important beginning, but successful cropping is a game of measurement and decision-making that aims to balance odds in favour of farmers. Technology and management skill are essential if farmers are to conquer the challenge of variable climatic and economic environments.

Towards 2050

There are great challenges ahead for cropping and crop production, and continuing need for new knowledge and technology. The world population that is now 7 billion is expected to rise to 9.2 billion by 2050. That introduces the challenge to increase current food production by 70 per cent because almost 1 billion people are currently underfed. Moreover, the challenge must be met on essentially the present cropping area of 1500 Mha because there is little opportunity for increase in the face of competition for nature conservation and other uses.

This large increase in crop production can only be achieved with combinations of greater crop yields and more intensive cropping adapted to local conditions and availability of inputs. Farming systems are dynamic and continuously adapting to changing ecological, environmental and social conditions, while achieving greater production and resource-use efficiency by application of science and technology.

The word sustainability is used loosely without useful definition. In the framework used in this essay, sustainable agriculture is defined as whether society is prepared to allocate enough resources to define problems, research them to find solutions in a reasonable time frame, and ensure adaptive remediation by the farmers.

The solution to "feed and green" the world in 2050 is to support this evolution more strongly by providing farmers with necessary information, inputs, and recognition. The challenge is, however, great. Irrigation, necessary to increase cropping intensity in many areas cannot be extended much more widely than at present, and it is uncertain if the current rate of crop yield increase can be maintained. Society needs greater recognition of food-supply problems, issues of food waste generally, and overconsumption in some societies. It must be prepared to increase funding and support for agricultural research to support the competent and innovative farmers in developed nations.

2

Cultivation Techniques of Crops

CULTIVATION METHODS

No crop is grown for hay alone; most of those described are used for sown pasture, or are field crops such as cereals or pulses. The cultural techniques and levels of input for other uses may differ from those for haymaking, and therefore that information should be sought elsewhere. Cultivation for hay is discussed, with the emphasis on techniques suited to small- and medium-scale farms. Hay crops are grown on arable land, where they must compete with other field crops on the grounds of profitability. As haymaking costs are proportional not only to the area worked, but also to the weight of the crop, intensive to semi-intensive methods aiming at high yields per unit area are therefore generally advocated.

Where land scarcity is the major limiting factor, increased yield per unit area may be the best way of improving fodder availability. This was demonstrated in Pakistan in an FAO-assisted project. Fodder crops cover some 19 per cent of Pakistan's total cropped area, but there is still a chronic forage deficit; the area under fodder remained static during 1977 - 1994, at around 2 600 000 ha, while yields rose over 20 per cent, from 17.4 t/ha to 21.4 t/ha, during the period. Well-grown crops of proven cultivars can produce 60 to 80 t/ha of green fodder (most fodder is irrigated), but there are still problems in assuring the supply of good seed and planting material at farm level.

In another study, in Punjab, Hanjra (1996) gives the results of 200 on-farm trials carried out over three years with the poorest and smallest dairy farmers in central Punjab.

The "improved" methods mostly had three parameters which differed from the farmers' own methods: seed quality; sowing method (including land preparation); and date of sowing. Yields are below the very high levels attained on research units and some large farms, but the effect of good seed, sown at the correct time, on a well prepared seed bed - all cheap inputs - is striking, with yields improved by 20 to 40 per cent, except for sorghum, where no improved fodder cultivars are available.

Table. Yields of fodders with traditional and improved cultivation in the Punjab

Crop	Yield (t/ha; green)		Percentage increase
	Traditional	Improved method	
Sorghum	18.5	20.0	7.5
Hybrid sorghum	-	52.5	-
Maize	15.5	19.4	20.1
Berseem	26.3	38.5	30.9
Lucerne	25.0	34.4	27.3
Oats	15.0	24.6	39.0
Berseem+sarson+oats	32.0	41.3	22.5

Note:

(1) Hybrid sorghum is multi-cut.

LAND PREPARATION

Clearing and levelling

Where fodder is cultivated for hay-making, the land must be prepared for mowing as well as put into a fit state for sowing and establishing the crop. Stones, stumps and termite mounds should be removed. Land with a long history of cultivation by the plough may require little or no treatment, but any field to be sown must be cleared of all obstructions to harvesting equipment.

Under rainfed conditions, land levelling to facilitate mowing, as well as to make sowing of small seeds at even depth possible, is often necessary: even small humps and hollows make sowing at a uniform, shallow depth impossible and lead to gaps and uneven stands. For irrigated crops, of course, proper levelling is essential for effective water distribution. In-field irrigation structures should be designed to facilitate harvest, especially for perennial and multi-cut crops.

A firm, level seed bed is necessary for all fodders for hay, and for small-seeded pasture-type crops it must also be fine. Preparation methods vary with the type of farm and soil conditions. Whatever the means of primary and secondary cultivations, rolling in the final stages of seed-bed preparation, where possible, is advisable for all small-seeded crops. In tropical and semi-arid areas, work should follow the contour and, where necessary, precautions against washout should be taken. Most fodders have excellent erosion control qualities, once established, but the time of sowing and a short while thereafter are critical.

Fertilizing

Hay is a demanding crop and mineral deficiencies must be made good before sowing. It is a highly extractive crop, since all is removed from the field. Farmyard manure is rarely returned to fodder fields in developing countries, although it may be used on other parts of the farm, especially for horticultural crops. In large areas of the irrigated subtropics where fodder cultivation is important, most of the dung is used as fuel. In mechanized, intensive, large-scale dairying, however, the excreta is now handled in aqueous solution - slurry - which is widely used for top-dressing grass, adding fertility (from concentrates and other bought in feeds) and disposing of a difficult by-product in an ecologically benign way.

Soil fertility status should be determined before pasture installation. Local knowledge may allow a reasonable estimate to be made, but analysis may be necessary. While levels of laboratory accuracy may be very high, the value of analyses are only as good as sampling accuracy, so care should be taken, and advice sought if necessary on sampling methodology.

Phosphorus (symbol: P) is a major nutrient for all crops and its value is well established for fodder and grassland. Soils are usually deficient in available phosphorus unless they have been heavily fertilized in the past. It is essential for root and seedling establishment and should be present in the seedbed in adequate quantity and in a readily available form. It is also very necessary for legume persistence, and phosphatic top-dressings are frequently necessary to support persistence. It is progressively fixed by soil minerals after application, and on some tropical soils this process may be rapid and severe and almost permanent. However, there is thereafter a slow release, on which the sward relies for development. In grazed swards, much of the phosphorus is recycled through droppings. In hay crops, of course, this is not the case.

Potassium (symbol: K) is an essential nutrient, with a relatively high concentration in green leaves. Hays contain 2 - 3 per cent in their DM, so it is a very potassium-extracting crop (a two-ton crop would remove the equivalent of 55 kg of potash fertilizer). It is required to complement nitrogen where high yields are the aim, as in hay production. Soil availability varies widely. Clovers are very sensitive to potassium deficiency.

Sulphur (symbol: S) is required in about the same quantity as phosphorus by plants, and legumes are often very responsive. However, it is much less used consciously as a fertilizer. Soil availability varies very widely. Some of the older fertilizers contained large amounts of sulphur; single superphosphate contains more sulphur than phosphorus; sulphate of ammonia supplies a lot of sulphur; basic slag, which was a major pasture fertilizer in industrial countries, contained appreciable amounts. More concentrated fertilizers with little sulphur are now common (partly to reduce transport and handling costs) and the steel-making process which produced slag is now little used, so the possibility of sulphur supply limiting production must be kept in mind.

Trace elements, or micronutrients, are limiting in some areas and application of tiny quantities in such cases can have spectacular effects. However, they should only be used, including cocktail mixtures, when a deficiency has been identified.

Nitrogen (symbol: N) is, of course, essential to all crop growth and its supply is one of the major limiting factors in forage production. An adequate nitrogen supply is essential, either as fertilizer or through biological fixation by legumes. Leguminous crops with effective nodulation will fix an appreciable amount of atmospheric nitrogen and add it to the overall farm nitrogen balance, but inputs of non-nitrogenous mineral fertilizer are needed to support this, as are increased management inputs in the case of mixed swards. For hay crops, heavy applications of N, including maintenance top-dressings between harvests, are limited by the resultant increasing density of sward, which hinders drying and can cause lodging and subsequent harvesting difficulties and losses. In small-scale farming situations, unbalanced use of fertilizer is a common fault: for unsophisticated users, the eye-catching responses to nitrogen top-dressing tend to eclipse the essential basal inputs of phosphorus and other necessary minerals. Nitrogenous fertilizer on legumes is wasteful and depresses biological nitrogen fixation.

Dressings for establishment will vary greatly according to local conditions, while maintenance fertilizer will depend on the management system. Levels should be decided according to local experience and research. A basal dressing of phosphatic (P) fertilizer is almost always necessary, and potash (K) is frequently needed; for non-legumes some nitrogen is necessary in the seed bed. Sometimes sulphur is deficient and leguminous crops, especially lucerne, can give large responses.

Calcium (symbol: Ca) is necessary for growth and legume nodulation. Under humid temperate conditions, heavy periodic liming, to raise the pH, is sometimes done. This is not suitable for tropical acid soils, but these may benefit from light dressings (600 kg/ha) of lime as a nutrient.

Micronutrient deficiencies should be dealt with where reported.

The forms of fertilizer used will often depend on local availability. For example, single superphosphate is recognized as being the best phosphate source for most fodders and pastures, but its phosphorus content is relatively low (17 to 18 per cent $P_2O_{5,}$ as opposed to 42 per cent in triple superphosphate), so it is not interesting if long transport is involved.

SEED AND PLANTING MATERIAL

The choice of crops for different situations and conditions. It is always advisable to use clean, healthy seed of the best adapted cultivars available. In developed countries, the pasture and fodder seed market poses few problems, although the availability of seed in bulk tends to be limited to a relatively narrow range of species and cultivars. In many developing countries, the fodder seed situation is difficult, especially for specialized fodder and pasture plants (as opposed to cereals or pulses used as fodders). Often, national priority has, in the past and probably correctly, been placed on producing seed of the major subsistence and cash crops, with fodder taking second place. The importance of livestock in most small-scale farming systems, and the present need for on-farm fodder production, however, mean that this imbalance should be rectified. The international seed market often provides only cultivars suited to a limited range of conditions (whatever the claims of the merchants) and is dominated by countries with a large internal fodder market, notably Australia, New Zealand and the USA.

Traditional fodder landraces, in developing countries, are often only available as farmers' seed from unselected crops, of very doubtful cleanness and purity. It has often been clearly demonstrated that fodder yields per unit area can be greatly increased, often more than doubled, by using good seed of locally proven cultivars and landraces.

Many countries have the knowledge and basic seed within their research systems; the problem is how to develop a sustainable, local commercial production and distribution chain. In the sections on individual crops, cultivars will be mentioned; it must, however, be kept in mind that although a cultivar has performed well under similar agro-ecological conditions on another continent or in another country, it does not mean that it can be transferred to a new area with confidence. For example, local disease patterns differ, therefore fodder sorghums selected abroad and imported to Africa, where there is a strong disease challenge, often perform very poorly.

IRRIGATED HAY

Fodder is a widespread irrigated crop in arid and semi-arid zones. Under small-scale farming conditions, usually within a mixed farming system, the greatest areas of cultivated fodder in the developing world are on irrigated lands in the Near East and the Indian subcontinent, often in wheat-rice farming systems and associated with milch buffaloes. Irrigation areas specifically for fodder are rare in the small-scale farming and pastoral sectors, but that in Xinjiang (Altai) is a notable exception.

In the large-scale sector, however, highly specialized fodder growing, again often associated with dairying, is common. Irrigation layouts should be so designed as to present the minimum hindrance to agricultural and haymaking equipment and should allow easy access for transport of the bulky crop. An adequate drainage system, to allow the fields to be dried off before mowing, is highly desirable.

Components of an irrigation system

Irrigation is not an end in itself, nor does the simple supply of water and seed to land assure a sustainable and economic increase in output. There is a complex interaction between the land and water base, agricultural practices and cropping choices, which interact on the system and affect its production, profitability and labour efficiency. The management and maintenance of irrigation require discipline; traditional systems depend on a stable community decision-making system, and all interventions require, well before proposal or execution, discussion at both household and community level. Irrigated farming communities are well aware of the systems' complexities. The main components are: (i) production activities (choice of crops, cropping and livestock system); (ii) on-farm water management (optimization of yield, and use of labour and inputs); (iii) the delivery system; (iv) the water supply system (relationship with local hydraulic system, reliability, quality, timing of supply); and (v) downstream implications (waterlogging, salinity, disposal of drainage).

Production activities

The choice of crops and cropping mixture is determined, within any climatic zone, by water availability, water quality, soils and the local market (both for sale of produce and supply of inputs). Where water is the limiting factor, the producers' aim is often to maximize return per unit of water.

On-farm water management

It is of the utmost importance that best use be made of the water delivered to the farm in order to optimize yield, not only of individual crops, but also of the farm crop mix, to optimize water-use efficiency, and to obviate land degradation due to waterlogging or salinization. The scheduling of water application to suit the plant-soil-water relationship of the site is essential. In addition to a good delivery system, in-field work is necessary. Proper land preparation, including levelling and grading, is essential to facilitate water application, economize on labour and power and to optimize input use (*e.g.*, watering must be so designed that fertilizer remains within the root-run of the crops and is not lost in the drainage water.

These operations are essential to the correct running of an irrigated farm; their application requires knowledge of crop agronomy, land and water management

The delivery system

The system between the offtake and the farmers' fields requires both management and maintenance. Management must assure equity in water apportionment and assure the timing of each farmer's water delivery. It must also ensure that the canals and works are properly maintained; prior agreement on how this will be done is a prerequisite to any interventions. Maintenance, usually through communal participation, is arduous and can be time-consuming. It is often possible, through some redesigning and modification, to provide systems that are less liable to siltation and therefore require less maintenance.

The supply system

Two main aspects must be studied: the effect of an offtake on water availability for other areas downstream, and possible negative effects on the hydrological system. The final decision on most major offtakes, therefore, cannot be decided by discussion with the potential users alone, as all parties likely to be affected must be consulted. Modern, solid structures not only greatly reduce maintenance labour but allow a modification and prolongation of the cropping system by permitting irrigation over a much longer season.

Downstream implications

Irrigation systems can have undesirable effects on land downstream through salinization, waterlogging and damage caused as a result of drainage water disposal. Both the design of the delivery and the management of the water at farm level are involved. It is very important that these aspects be taken into account before interventions are undertaken. Since the user community may not be directly affected by the downstream damage, such effects must be judged in the light of their wider community and environmental effect.

TYPES OF IRRIGATION

Surface irrigation, in which the water is distributed through flooding by basin, border and furrow, is the ancient, traditional system, which still accounts for the vast majority of the world's irrigated farmland. It is likely that most irrigated small-scale farming fodder will be watered in this way. Surface water is distributed in several ways, and the main ones are discussed briefly below insofar as they affect fodder work.

The subject is very complicated. Standard texts should be consulted for information on layout of irrigation units, their organization, and details of irrigation methodology. Crop water requirements and guidelines for their prediction are discussed in another FAO publication by Doorenbos. In considering irrigation of hay, however, care should be taken that the method chosen does not unduly hinder the use of whatever machinery is foreseen and that there will be adequate access at haymaking time for the type of transport to be used to remove the crop.

Basin irrigation

Basin irrigation is the oldest and simplest system: a field is divided into one or more basins; each basin is a piece of level land surrounded by a bund which retains water until it has soaked into the soil. It can be adapted to many crops and farming systems, but for fodder production it is very labour demanding and practically precludes any mechanization of harvesting, unless very large basins can be used. The in-field bunds have to be removed at each cutting, and in the case of multi-cut crops, reconstructed thereafter.

Border irrigation

Border irrigation (also known as border-check and border strips) resembles basin irrigation in that the fields are divided into units by bunds, but borders slope away from the farm channel in the direction of water flow. They are not level and the method of irrigating is different. The water is not retained on the field to soak in, but soaks in as it flows over the land, so it is important to use the correct flow for the correct duration to ensure that the correct amount of water infiltrates into the soil. This system is much better suited to haymaking than is basin irrigation. If cultivation and/or harvest is mechanized, border width should fit the equipment to be used.

Furrow irrigation

This is the most widely used system for row-crops: water does not flow over the entire soil area, but is confined to furrows between the crop rows, and infiltrates into the bottom and sides of the furrow. It is excellent for large fodder crops such as Napier grass, maize and sorghum, but it is not suited to hay corps. If cultivation and/or harvest is mechanized, row width should suit the equipment to be used.

Spate irrigation

Here the supply is dependent on torrents with only occasional flow, as after heavy rain or snow-melt. Traditional offtakes are very labour intensive, can often only be used when river levels are low, and may be completely destroyed by snow-melt floods in summer at a season when irrigated cropping can be very productive.

Qanat

This is a system (also called *karez*) of tapping underground springs by gently sloping tunnelling to transport it (underground) to cultivated land. It is widespread in parts of Iran and Afghanistan and is used as far east as the Turfan Depression in China.

Piped systems

Various systems of (usually underground) piping of water have been devised to reduce the enormous wastage of open-channel systems and to thus free more land for cultivation.

The water is usually supplied to the field through up-pipes with valves (risers) in the case of underground pipes, or through gated pipes where they are on the surface. Accurate levelling and some in-field structures are still required for such systems. Alfalfa risers have their valves just below mean soil level so that once the crop is established the mower and other machinery can pass over them unobstructed.

Sprinkler irrigation

This requires a considerable investment in equipment, but has the advantage that great investment in land-levelling, drainage and other infrastructures is not needed; less land is occupied with irrigation structures; water is more efficiently used; and there is minimal danger of seepage losses and salinization through rising water tables. It does away with the need for in-field structures which hinder fodder harvesting.

Centre pivot

Centre pivot systems are large, computer controlled sprinkler systems which rotate around a central axis. They are widely used for fodder crops in semi-arid or desert areas, such as Saudi Arabia and California. They are primarily for large-scale commercial farms where fodder is a high-value crop, and are well suited to the task.

Cold season watering of hay-fields

Watering of natural hay fields before the thaw is traditional in parts of Russia and Mongolia. Water is diverted from springs and streams on to areas reserved for hay, with or without some in-field structures, and is stored by freezing on the surface. Traditional water-spreading methods practised in the Mongolian mountain-and-steppe zone involve sporadic diversion of streams to develop ice-sheets over hay land which will, thereafter, melt at the onset of the growing season.

TECHNIQUES OF CROP PRODUCTION

Crop improvement technologies improve the potentialities of higher yield and food quality in crops. New varieties and hybrids of various crops released by various crop improvement techniques are ultimately aimed to increase the yield and better quality.

But, these potentialities are realisable only when crops are managed and grown under conditions conducive to express their genetic potentialities. The techniques to release the potentialities of varieties and hybrids of various crops are termed as crop production techniques.

These techniques include:

1. Soil management techniques - which in turn include soil fertility management practices and soil conservation practices.
2. Water management techniques - both under irrigated and rain fed situations.
3. Weed management techniques - which include chemical weed control, cultural weed control and integrated weed control practices.
4. Cultivation practices to grow crops.
5. Nutrient management techniques.

All these aim at providing most congenial situations for a crop to express its potentiality. Among them, nutrient management techniques aim at providing sufficient nutrients at the required stage in the required form, while water management techniques aim at providing ideal situations for growth of crops. Weed management techniques aim at keeping the land weed free so that crop does not experience competition by unwanted plants, while soil management techniques aim at maintaining the fertility of soil and retention of soil in such a way that it is available for long term production.

Nutrients can be provided to crops by fertilizers or manures or both. As the crop plants remove large quantity of nutrients for their growth, these nutrients need to be replenished in such a way that they are available to crops on long term basis.

IMPROVED METHODS OF CULTIVATION

Agriculture is the backbone of Indian Economy. With the increasing need of population the traditional trend of agriculture is changing day by day and becoming a business-oriented sector.

To cope with the growing population improved techniques of cultivation have developed.

IMPROVED METHODS OF PROPAGATION

Mciropropagtion

Mciropropagtion involves the production of plants from very small plant parts, tissues or cells grown aseptically in a test tube or other container where the environment and nutrition can be rigidly controlled. The growing of plant tissue such as callus, cell suspensions and various plant organs (stems, flowers, roots and embryos) in artifical meidum is known as tissue culture. Sometimes, tissue culture is synonymous for micropropagation.

The main advantage of this technique are rapid mass propagation of clones and can produce disease free seedlings.

SEEDS AND SOWING TECHNIQUE

Establishment of a good stand is the pre requisite for attaining high yields. It depends upon time, depth and method of sowing. Use of quality and improved seed play a role in increasing productivity. Seeds are subjected to special treatement like scarification, dipping in hot water, hormonal treatment etc. To reduce attack from these diseases, the seeds are treated with fungicides like thriam, captan, mancozeb, carbendazim, captfol etc.

Optimum time of sowing

Sowing the crop at optimum time increases yields due to suitable environment at all the growth stages of the crop. Early flowering is induced and moisture stress or dry spells may be avoided during critical stages. The optimum time of sowing for most of tropical crops is immediately after the onset of monsoon *i.e.* June or July. The optimum time of sowing for temperate crops like wheat and barley are from last week of October to first week of November. The optimum time of sowing for most of the summer crops is first fortnight of January.

Depth of sowing

For obtaining good stand of the crop,it is essential to sow the crop at optimum depth. Crops with bigger sized seeds like groundnut, castor, sunflower etc. can be sown even upto a depth of 6 cm. Small sized seeds like tobacco, ragi, have to be sown as shallow as possible.The optimum depth of sowing for most of the field crops ranges between 3 cm to 5 cm.

Method of sowing

Drilling in lines with ferti-cum seed drill has come in practise.also transplanting is followed in some crop to obtain better crop stand. The poly bag culture technique is now practised for rice, Polymulching is followed in groundnut.

USE OF HIGH YIELDING VARIETIES

To increase the production, high yielding hybrid varieties with short duration are to be used. A considerable efforts have been taken to incorporate resistance and quality into the new variety,*e.g.* eight sorghum hybrids are developed among which CSH-5, CSH-4 are resistant to downy mildew and CSH-5, CSH-6 are resistant to shootfly.

CROPPING SYSTEMS AND ROTATIONS

Depending on the resources and technology available, different types of cropping systems are adopted on farms. It includes intercropping, mixed cropping and sequences cropping. Intensive method of cropping is multiple cropping in which more number of crops within a year and more number of crops on the same piece of land are cultivated at any given period.

Sorghum-soybean and Sorghum- Groundnut have remunerative intercropping systems.

INTERCULTURAL OPERATIONS

Gap filling

In long duration and widely spaced crops gap filling os to be followed which helps to control the weed and increase yield.

Thinning

Thinning is needed to obtain optimum plant population by removing excess population.

Use of balance fertilizer

Now a days use of technical grade fertilizers with optimum dose are used. Suitable methods of fertilizer of application are used depending upon physical form, solubility, and mobility.

Liquid fertilizers are applied with irrigation water. Fertilizers are applied by soil placement, fertigation etc. Demand for use of biofertilizers like rhizobium, blue green algae is increasing.

Use of improved method of irrigation

Apart from use of traditional methods of Irrigation like flooding, border, check basin now day methods like drip, sprinklers are in use to save water and offer several advantages.

Plant protection measures

Integrated pest management, use of bio pesticides, use of technical grade synthetic chemicals, bio control measures is important.

HARVESTING AND STORAGE TECHNIQUE

Post harvest operations are assuming importance due to higher yields and increased cropping intensity. Due to introduction of modern technology, yield levels have substantially increased resulting in marketable surplus which has to be stored till prices are favourable for sale.

Time for harvesting can be known approximately by the duration of crop. As maturity depends on climate, maturity symptoms are good indicators for deciding the time for harvesting. Olpad thresher is used for wheat, barley oats etc. Japanese paddy thresher is used for threshing. Artificial methods like steam drying are in use.

CROP PRODUCTION TECHNOLOGY

Cotton is one of the most important commercial crops playing a key role in economic and social affairs of the world. It is an important fibre crop, which is cultivated in more than 80 countries of the world but ten countries viz USA, CIS, China, India, Brazil, Pakistan, Turkey, Mexico, Egypt and Sudan account for nearly 85 per cent of the total production.

There are four cultivated species of cotton, *viz.* Gossypium hirsutum, G.barbadense, G.arboreum and G.herbaceum. India is the only country in the world where all the four cultivated species are grown on commercial scale, besides hybrids. The maximum area is covered by hybrids (40,D/o), followed by G.hirsutum (36,Yo), G.arboreum (16,Yo) and G.herbaceum(8'Yo). The area under G.barbadense is negligible (0.2'Yo). In India, there are nine major cotton growing states which are divided into three zones *viz.* North Zone (Punjab, Haryana and Rajasthan), Central Zone (Maharashtra, Madhya Pradesh and Gujarat), and South Zone (Andhra Pradesh, Karnataka and Tamil Nadu). Nearly 65 per cent of the cotton crop is grown under rainfed conditions.

Soil Requirement

Cotton is grown on a variety of soils ranging from well drained deep alluvial soils in the north to black clayey soils of varying depth in central region and in black and mixed black and red soils in south zone. Cotton is senii-tolerant to salinity and sensitive to water logging and thus prefers well- drained soils.

Crop Season

The sowing season of cotton varies considerably from tract to tract and is generally early (April-May) in northern India and is delayed as we proceed to down south. Cotton is a kharif crop in the major parts of the country *viz.* Punjab, Haryana, Rajasthan, Uttar Pradesh, Madhya Pradesh, Gujarat, Maharashtra and parts of Andhra Pradesh and Karnataka.

In these areas, the irrigated crop is sown from March-May and the rainfed crop in June-July with the commencement of the monsoon. The pre-monsoon dry sowings, practiced in parts of Gujarat and Madhya Pradesh in the end of May or in early June have been found to give an early start with the resultant increased yields.

In Tamil Nadu, the major portion of the irrigated and rainfed crop is planted in September-October, whereas the sowing of the rainfed crop in the southern districts is extended upto November. In parts of Karnataka and Andhra Pradesh, the desi cotton is usually sown in August-September. Early sowings done in the Dharwad and Gadag area in the beginning of September have been found to give better yields.

In addition, summer sowings in Tamil Nadu are done during February-March. The sowings of cotton in the rice fallows of Andhra Pradesh and Tamil Nadu extend from the second half of December to the middle of January.

Cotton based cropping system:

The type of cropping system adopted *i.e.* mono-cropping, mixed cropping, inter-cropping and rotation/ sequence cropping is dependent upon the amount and distribution of rainfall, length of growing season and type of soil.

The common traditional practice of cotton cultivation in many parts of central and south India is strip cropping of 1 or 2 rows of pigeonpea after every 8-10 rows of cotton, 3-5 rows of ragi Otemated with 1 row of cotton.

Table. Dominant cotton based cropping systems in different states

States	Cropping Systems
Punjab, Haryana and Rajasthan	Cotton-Wheat; Cotton Mustard
Madhya Pradesh, Maharashtra , Guajarat,	Monocropped cotton, cotton-jowar (two and year rotation), intercropping with black gram, green gram, soyabean, groundnut and red gram.
Andhra Pradesh	Monocropped cotton, Cotton-Rice (sequence), Cotton-Chilli or Cotton- Tobacco(two year rotati6ns)
Tarnil Nadu	Monocropped cotton, Rice-Cotton, Rice- Rice-Cotton. Cotton- jowar, Cotton-Pulse- Jowar, intercropping with onion, groundnut and blackgrarn.
Kamanataka	Monocrapped cotton, Cotton-Wheat, intercropping with chilli, groundnut, blackgram, green gram.

Preparation of land

The time available for land preparation following the harvest of wheat is limited in the north one.The field is irrigated after the harvest of wheat and ploughed. Thereafter a light irrigation is provided and the field is ploughed and the soil is planked with a wooden plank.

In central and southern zone of India where cotton is a rainfed crop, deep ploughing is recommended to destroy perennial weeds once in 4-5 years.. The field is prepared by repeated ploughing with a blade harrow with the onset of pre- monsoon rains.

In red soils of southern zone having poor moisture retention, field is given 2-3 light ploughings after harvesting.

The availability of herbicides for weed management has opened up options for reduced tillage and residue management which are more energy efficient and have an ameliorative effect on soil quality.

Selection of varieties and hybrids

Several high yielding varieties/ hybrids of cotton have been reconnnended and released for commercial cultivation in different states. Varieties of G.hirsutum and G.arboreum are grown in all the cotton growing states. G.herbaceum varieties are cultivated in some areas in Gujarat and Karnataka. G.barbadense is cultivated on small area in Tamil Nadu and Karnataka states. The cotton hybrids are cultivated in the central and south zones. Some hybrids have been released recently for northern zone.

Method and time of sowing

Cotton is sown using tractor or bullock drawn seed drill or by dibbling. Hand dibbling of seeds at recommended spacings is commonly practiced in rainfed- areas particularly for hybrids. This system ensures proper plant stand, uniform geometry and also saves seeds. Later, the crop is thinned to the recommended population.

Time of sowing: Sowing time for cotton varies considerably in different cotton growing zones.

Seed Rate, Spacing and Plant Population

The plant population and geometry varies. with the plant architecture, soil f@rwity status, soil moisture holding capacity and species/ hybrid of cotton. Closer spacing is recommended for dwarf compact short duration varieties..

Nutrient Management

Climatic factors like rainfall and temperature, soil factors like soil type, depth, pH, EC, CaCO3 content and organic matter, crop factors like sequence, and plant factors like genotype and duration determine the nutrient requirement of cotton. Organic manure is added once in 2-3 years in central and southern zones but this is seldom-practiced in northern zone. Integrated nutrient management schedules with a part of chemical fertilizers supplemented by farm wastes (including FYM), crop residues and bio-fertilizers and green manure are also being worked out. The fertilizer doses and method of application recommended for cotton in different zones.

Inter cultivation and weed control

Inter-cultivation is done fairly regularly by either a blade harrow with a three lined hoe or a desi plough. In the crop sown by broadcasting, one or two hand hoeings are given to remove weeds, inter-cultivation not only checks the growth of weeds but also leads to better soil aeration and soil-moisture conservation.

Weeds compete with cotton crop for nutrients, light and moisture. Cotton is susceptible to weed competition from sowing to about 70 days when the canopy covers the inter- spaces. Cotton yields are reduced by 50 to 8511/o if weed growth is unchecked. Fluchloralin or pendimethalin @ 1 kg ai/ha. as pre-plant incorporation with one hand weeding and crosswise hoeing has been recommended for satisfactory weed control. Deep rooted perennial weeds are removed by summer ploughing.

The thinning of the cotton crop is a special feature of the irrigated crop sown on ridges in Peninsular India and in parts of southern Gujarat. Thinning is desirable for maintaining the optimum population of plants to obtain a high yield. During thi,nning, the vigorous seedlings are retained and weak seedlings and off types are removed.

Water management

Depending upon the climate and crop growing period, cotton needs 700-

1200 mm water to me6t its maximum water requirement. The water requirement is low during the first 60-70 days after sowing and is highest during flowering and boll development stages.

Irrigation Scheduling

Cotton is commonly flood irrigated although irrigation by furrow or alternate furrow method is more effective and water saving. Drip irrigation is becoming popular particularly in the hybrids for central and southern zones. Cotton needs, to be irrigated at 50-70 per cent depletion of available soil moisture. Based on climatological criteria of scheduling irrigation of 0.6 IW/CPE with 6 cm depth is commonly recommended. On sandy loam soils of northern zone 3-5 irrigations are commonly given. On red sandy loam soils of Tamil Nadu with low water retention capacity, 4-13 light irrigations may be necessary.

Rain Water Management

Rainfed cotton yields are low owing to erratic and uneven distribution" of rainfall. Rainfed cotton suffers from moisture stress during post monsoon season which coincides with flowering and boll development stages. The rainfed crop depends on residual soil moisture and several measures have been recommended to conserve rainwater in situ or harvest excess run-off and recycle it at critical phases. Cotton cultivation on ridges across the slopes conserves more water, reduces soil erosion and improves yield.

Ridge sowing on 0.4 per cent slope and tieing ridges at 6.0 m interval just before the normal withdrawal of monsoon is recommended.

Collection of excess water into farm ponds and recycling it to provide one life saving irrigation at the critical early boll development phase also improves rainfed cotton yields.

Insect pest and disease management

Cotton is attacked by a number of insert pests. The important ones are jassids, aphids, thrips, whiteflies and bollworms (Pink, spotted and American) and Spodoptera. For the the management of sucking Pests Methly demeton @ 2 MVI should be sprayed.For the management of whiteflies neem product @ 2.5 1/ha) or neern seed kernel extracts (5 per cent) may be sprayed. In case of severe infestation,Triaophos @, 600-800 ml/ha should be sprayed. During flowering period, Fndosulfan, chiolpyriphos or Quinalphos (2.5 l/ha) should be sprayed for the management of bollworms. Moreover, the insecticides should be sprayed only when the population of insect pests is beyond economic threshold level.

The insecticides should be sprayed as per the recommended doses. Insecticide resistance has been observed in some insects, hence proper selection of insecticides is must for managing the insect pests. HNPV should be used to manage Helicoverpa during early stages. Synthetic pyrethroids should not be used more than twice and be alternated with other conventional insecticides. As the resistance level to synthetic pyrethroids in American bollworm is very high, their use should be avoided during peak Helicoverpa infestation. However, t he synthetic pyrethroids are still effective against pink and spotted bollworms.

Cotton suffers from a number of diseases. The important ones are bacterial blight, leaf spots caused by Alternaria and M.yrothecium, grey mildew, boll rots, Fusarium wilt, Vejtcillium wilt, root rot and leaf curl virus. In the areas where the bacterial blight appears in severe form acid delinted seeds should be treated with Streptocycline (0.01 per cent) before sowing. Later a foliar spray of a combination of Streptocycline (0.0111/0) and Copperoxychloride (0.25 per cent) should be given when the disease starts appearing. Copperoxychloride is effective against fungal leaf spots too. Grey mildew is managed with the use of wettable sulphur (0.2To) or carbendazim (0.1 per cent).

For the management of boll rots, Copperoxychloride or carbendazim should be sprayed along with the insecticides. Fusarium and Verticillium wilts are best managed by growing recommended resistant varieties in the wilt affected areas.

In an area where root rot occurs regularly, seeds should be treated with carbendazim @ 2-3 g/kg seed before sowing. When the root rot symptoms start appearing, drenching of affected plant as well as nearby plant with 0.1 per cent carbendazim should be done. Intercropping with mothbean is also recommended in such area. Application of Zinc sulphate @ 24 kg/ha. also helps in reducing the diseases. Seed treatment and soil application of some antagonists like Trichogramma viride, Thrzianum, Gliocladium virens and Pseudomonas fluroescenes have shown promise under experimental conditions.

Leal curl disease caused by Gemini virus has become prominent in the northern cotton zone. The virus is transmitted by whitefly. Therefore, the crop should be free from the whitefly attack. The insecticides, which control whitfly should be sprayed. Desi cottons are not affected by this disease. Some var varieties like LRK-516, LRA-5166, RS-875,LHH-144 are resistant to this disease. These varities should invariably be grown in leaf curl prone areas.

Picking and Marketing

Cotton should be picked after boll bursting, during sunny days to ensure moisture free kapas of good quality. Delay in picking leads to falling of Kapas on the ground which results in deterioration of quality due to mixing of soil articles, leaf bits and other trash. High temperature is also detrimental. The picked kapas should be properly cleaned before taking to the market for sale.

There is a monopoly cotton procurement scheme in Maharashtra while in all other states, the kapas is sold in the open market. The produce should be graded to ensure proper price.

Technologies recommended during the last one year.

(i) Varieties and hybrids recommended during the last one year

Region / State	Production condition	Varieties / Hybrids	Special feature
Andhra Pradesh	Assured rainfall and irrigated tracts of A.P	RCH-2 Intra hirsatum hybrid	----
North Zone	For the entire North Zone as an irrigated crop;	AAH-1 First intra arboreum hybrid.	----
Haryana	Irrigated tracts of Haryana	HD-123 G.arboreum variety	

(ii) Varieties and hybrids identified for release

Region/State	Production condition	Varieties/Hybrids	Special feature
Rajasthan	Recommended for irrigated agro ecosystem and for cotton wheat rotation	RS-810	Resistant to Cotton leaf curl virus
Rajasthan	Recommended for Rajasthan	RG-18	Better fibre properties
Punjab	Recommencied for iadaptive trial in irrigated agro system of Punjab State	LDI-694	Resistant to Cottonleaf curt virus
Haryana	Recommended for Rajasthan irrigated agro-eco-system of North Zone	AAH-1	Short stapled, early maturing, GMS based intraarboreum hybrid.
South zone (TN, AP, Karnataka)	Recommended for both rainfed and irrigated conditions.	Sumangala.	Medium staple variety with good fibre maturity and fibre elongation.
Tamil Nadu	Recommended for rainfed tracts of TN	SVPR 3	--
Tamil Nadu	Recommended for the winter irrigated combodia tracts.	MCU-12	--
Andhra Pradesh	Recommended for	LAHH-4	Medium staple

	entire Andlua Pradesh		cotton hybrid.
Andhra Pradesh	Recommended for entire Andhra Pradesh	L-604	Medium staple cotton having tolerance to bollworm
Andhra Pradesh	Recommended for the dry farming tracts of Rayalseema of AP suitable for double cropping of cotton and groundnut in irrigated conditions.	Aravinda	
Karnataka	Recommended for Raichur belt.	Sabana	Medium staple cotton with tolerance to drought and bollworm.
Karnataka	Recommended for irrigated tracts.	RAMPBS-155	--
Maharashtra.	Recommended for Marathwada region	PHH-316	Medium staple cotton with tolerance to drought and sucking pests
Gujarat	Recommended for Saurashtra and Kutch region.	Vikrama	Medium staple hybrid with an excellent fibre maturity and strength.
Andhra Pradesh	Recommended for southern Andhra Pradesh	ICMF-20	--

Production and Protection technologies recommended during last one year.

Agronomy

Cowpea and Soyabean as an intercrop with cotton has been found to be beneficial. Apart from increasing monetary returns to the farmers, the intercrops increase the predator activity in the field and thereby decrease the sucking pest population. Cowpea as a live mulch was also found to be better than Sunhemp in soil moisture conservation under rainfed situations in the Central Zone. Foliar application of 1 per cent MgSO4 twice on 60th and 80th day was also found to increase seed cotton yield.

Entomology

1. Seed treatment with imidacloprid (Gauho) @ 7.5 g/kg dose as chemical was found very effective in controlling sucking pest upto initial 60 days.

2. New insecticides molecules - Spinosad 48 EC @ 50-75 g.a.i/ha. Indoxacarb 15 EC at 75 g.a.i./ha. and Decis tablet 25 per cent @ 12.5 g.a.i./ha. were consistently effective for control of bollworms in cotton.
3. Insect growth regulator Rimon 10 EC (Novuluron) @ 100 g.a.i./ha. and Match (Lufenuron) 5 per cent EC @ 60 g.a.i./ha. were highly effective for control of bollworms in cotton.
   Sprayable formulation of new insecticides Acetamiprid 20 SP @ 15 g.a.i./ha. Confidor 200 SL and Thiomethoxam 25 g.a.i./ha. were good for control of sucking pests namely aphids and jassids. Thimethoxam 25 WG at higher dose of 50 g.a.i./ha. was effective for control of whiteflies.
4. The important strategy of IRM is to use insecticides from more than one group *viz.*
   - In rotation or alternation of groups for
   - As mixtures or
   - In a mosaic pattern or
   - Imposition of a temporal restriction on the use of a particular group of insecticide.

This is achieved by strategic aimed at curing resistance or preventing the development of resistance to existing or new pesticides. There is no single strategy that can be applicable universality for all situations. IRM strategies have to be formulated necessarily based on existing situations. The seed treatment with carbosulfan @ 20 g/kg. of seed or imidacloprid @ 7.5 gm/kg of seed was introduced as a new protection technology for the north zone.

Critical Technologies which are recommend earlier but need. special attention

I. Potash application to hybrids under high yielding situations.
II. Need to correct secondary and micro-nutrient deficiencies, particularly Mg., Zn and Boron.
III. Drip irrigation wherever possible can be adopted to save water and increase the efficiency of fertilizer input.
IV. Fine tuning of Integrated Pest Management for specific locations. New and safer chemicals like Imadacloprid and Rimon and Spinosad need to be incorporated in the IPM module to reduce environment pollution and health hazards.
V. Cotton Leaf Curl Virus (CLCUV) disease caused by a Gemini virus and transmitted by whitefly has been a concern to farmers of north zone since last 4-5 years. This year also the disease has appeared in the second week of May. Extensive surveys in Haryana, Punjab and some parts of Rajasthan have been carried out and the incidence is low, but the disease is present in all the areas. In some villages however, like Chautala, Saktakhera and jandwala the disease incidence

upto 40 per cent in isolated fields has also been noticed. The following recommendations have been made for the management of this disease.

- Uprooting and destruction of CLCUV affected plants in those fields where only few disease plants are observed.
- Eradication of weeds harbour-ing CLCUV like AL utilon sps (Peeli buti), sida sps (kanghi buti), Xanthium strurnarium (Gut patna). Achyronthes aspera (Puthkanda) and Ageratum canyzoides around the fields and water channels is advocated.
- The whitefly incidence has been low in the area. However, the control of whitefly with insecticides is essential to check the further spread of disease. Farmers are advised to spray their crop with any of the following insecticides for the control of whitefly.

S.No.	Insecticide	Dose/acre in 200 litre of water
1	Oxydemetan methyl (metasystox) 25 EC	400 ml.
2	Dimethoate (Rogor) 30 EC	350 ml.
3	Triazophos (Hostathion) 40 EC	600 ml.
4	Ethion (Formite) 50 EC	800 ml.
5.	Acephate (Starthene) 75 SP	400 g.

IPM/ IRM MODULES

North Zone

1. Use of tolerant variety
2. Seed treatment with Imidacloprid @ 5 g/kg seed
3. Hand picking of eggs/larvae and damaged plant parts.
4. Weekly monitoring of key pests.
5. Release of parasite/ predators - Trichogramma sp. Chrysoperla sp.
6. Neem @ 1.5-2.5 litre/ha.
7. Endosulfan @ 700 g.a.i./ha.
8. Fenvalerate @ 50-75 g.a.i./ha.
9. Ha NPV @ 450 LE/ha.
10. Triazophos @ 500-600 g.a.i./ha. (whitefly above ETL).

Central Zone

1. Seed treatment with Imidachloprid @ 10 g/kg. of seeds for sucking pests.
2. Early release of Chrysoperla @ 10,000 larvae/ ha.
3. Spraying of neem formulation @ 50/o.
4. Release of Trichogramma @ 1.5 lakh/ha. (Twice)

5. Spraying of Endosulfan 0.07'Yo.
6. Spraying of HNPV @ 450 LE/ha. for Heliothis.
7. Spraying of Neem formulation @ 5'Yo
8. Hand collection of the full grown larvae as and when required.

South Zone

1. Seed treatment with lmdiacloprid (10 g/kg seed).
2. Application of 50/o NSKE.
3. Release of egg parasitoids (7@icho,-rama @ 1.5 lakh and Chrysopa @ 0.5 lakh/ha. between 45-75 days after sowmg.
4. Pheromonetrapsshouldbeerectedtoattractmalemoths of bollworms.
5. Endosulfan 35 EC @ 800 mvacre
6. Hand collection of grown up larvae and bollworm infested fruiting bodies at regular intervals.
7. Topping of plants should be done to reduce scope of egg laying by bollworms.

Insecticide Resistance Management

- Cotton occupies only 50/o of the total cultivable land but consumes more than 550/o of the insecticides used in the industry.
- Insecticide resistance in the cotton bollworm and other pests leads to poor pest control and consequently repeated and indiscriminate application of insecticides. Resistance problem in the cotton bollworm Helicoverpa armigera resulted in a pesticide 'treadmill', poor yields, excessive expenditure and thus financial losses thereby leading to farmer suicides in some parts of India in 1987 and 1997.
- The current strategies are based on results of a network project on 'Insecticide Resistance Management of Helicoverpa armigera'carried out at six different research institutions in India. The strategies place emphasis on efficient use of insecticides to conserve the ecosystem for better pest management. These have been extensively tested for 4-5 years in hundreds of acres in Maharashtra, Andhra Pradesh, Taniil Nadu and Punjab and were found to reduce insecticide use by 50-90 per cent with yield increases of 10-25 per cent.

Strategies to manage 'Insecticide resistance in cotton pests'

- Ensure insecticide use such that Resistance is low, Natural enemy populations are least disturbed and different groups of chemicals are alternated and
- Use host plant resistance for sucking pests to avoid insecticide sprays early in the season. Cultivate 'sucking pest tolerant' cultivars.
- Zero insecticide - till 60 days.

- Based on Economic thresholds the following simple 'window strategy' can be adopted.
- No 'organophosphates' till 90 days.
- Ensdosulfan - not beyond 90 days.
- Biorational eg. HaNPV, neem etc. 'if available' only at 80-90 days.
- Pyrethroids only after 110 days.

Days after sowing				
0-60	**60-90 .**	**80-90**	**90-110**	**110-130**
Zero spray	Endosulfan	Biosprays	Organo-Pyrethroid	phosphate

Sucking pest resistant genotypes

- Help in delaying the first spray, thus conserving the initial build-up of natural enemies.
- Damage due to jassids or aphids in tolerant genotypes does not have any significant negative impact on yields.
- A certain extent of jassid damage to plants was also found to discourage infestation of Helicoverpa armigera.

Table. Selection of Cultivar for Central and South India

Recommended hybrids :	NHH-44, PKV HY-2, JK Hy@l, IK Hy-2, H-8, H-10, Ankur-651
Recommended varieties:	LRA-5166,LRK-516,PKV-081. Desi types AKH-4, AKH-8401

Selection of Cultivar for North India

Early maturing	F-846, F-1378, LH-1556, F-2054, H-1098
Recommended hybrids	Omshankar, Fateh, Dhanalaxmi
For CLCV prone areas	RS-875, LHH-144, LRK-5i6, LRA- 51ra6, All desi types (LD-327, HD-107, LD-491).
Resistance/ tolerance to jassids	Bikaneri Narma, H-777, RS-875, RST-9, F-5-5, Fateh.

Seed treatment

Imidacloprid 7 g Carbosulfan 50 g/ kc. seed helps in delaying the first spray for hybrids in protecting the crop against jassids upto 40-60 days.

Economic thresholds for Helicaverpa

- Based on our studies, we suggest Economic thresholds of 10 larvae/ 20 plants during early fruiting phase when squares are predominant and later 20 larvae per 20 plants sampled in one acre, when boR formation is at its peak.
- Larval and egg threshold scouting is cumbersome and individual farmer dependent.

- Instead of searching for larvae, farmers can score plants as having a larvae if it has 2-3 flared up squares on a sympodia.

Endosulfan

- Fndosulfan is relatively less toxic to natural enemies.
- Endosulfan may be used as an emergency option on jassids at an ETL of 2/ leaf.
- Endosulfan should be encouraged only as early season spray against Helicoverpa at an ETL of 50 per cent damaged plants.
- Resistance levels have been found to be invariably low early in the season *i.e.* August - September and higher later on.

Organophosphates and Carbamates

- Avoid use of broad spectrum 'OP's such as monocrotophos, metasystox, dimethoate, acepl.z.te, phosphoniidon etc. especially as early season sprays as these strongly disrupt the natural enemy populations.
- Resistance levels against certain organophosphate group of insecticides (Quinalphos, Chlorpyriphos and Profenophs) and carbamates (Thiodicarb) have been found to be low in most populations tested.
- Organophosphate's can be used as effective larvicides during 90-110 DAS based on ETLs I 00 per cent affected plants.

Pyrethroids

- Pyrethroid resistance is high in many part of India.
- But, these can be effective on Helicoverpa only on younger larval stages or adults or if used along with synergists such as sesamum oil or organophosphates such as ethion or profenophos.
- Pyrethroids are still effective on spotted and pink bollworm.
- Hence pyrethroids (Cypermethrin, fenvalerate or Karate) in Central South India, can be used for late season (Oct- Nov) sprays to be used with Ops (Profenophs, Ethion, Quinalphos or Chlorpyriphos) to combat resistant Helicoverpa and pink bollworm incidence (at an ETL of 8 moths/ trap/ night for three consecutive nights). Avoid ransom mixtures.

3

Nutritional Management of Crops

NUTRIENT MANAGEMENT

Nutrient management is the science and art directed to link soil, crop, weather, and hydrologic factors with cultural, irrigation, and soil and water conservation practices to achieve the goals of optimizing nutrient use efficiency, yields, crop quality, and economic returns, while reducing off-site transport of nutrients that may impact the environment. Nutrient management is the skillful task of matching a specific field soil, climate, and crop management conditions to rate, source, timing, and place (commonly known as the 4R nutrient stewardship) of nutrient application.

Fig. Nitrogen fertilizer being applied to growing corn (maize) in a contoured,no-tilled field in Iowa.

Some important factors that need to be considered when managing nutrients include (a) the application of nutrients considering the achievable optimum yields and, in some cases, crop quality; (b) the management, application, and timing of nutrients using a budget based on all sources and sinks active at the site; and (c) the management of soil, water, and crop to minimize the off-site transport of nutrients from nutrient leaching out of the root zone, surface run-off, and volatilization (or other gas exchanges).

There can be potential interactions because of differences in nutrient pathways and dynamics. For instance, practices that reduce the off-site surface transport of a given nutrient may increase the leaching losses of other nutrients. These complex dynamics present nutrient managers the difficult task of integrating soil, crop, weather, hydrology, and management practices to achieve the best balance for maximizing profit while contributing to the conservation of our biosphere.

NUTRIENT MANAGEMENT PLAN

A crop nutrient management plan is a tool that farmers can use to increase the efficiency of all the nutrient sources a crop uses while reducing production and environmental risk, ultimately increasing profit. It is generally agreed that there are ten fundamental components of a Crop Nutrient Management Plan. Each component is critical to helping analyze each field and improve nutrient efficiency for the crops grown. These components include:

- *Field Map:* The map, including general reference points (such as streams, residences, wellheads etc.), number of acres, and soil types is the base for the rest of the plan.
- *Soil Test:* How much of each nutrient (N-P-K and other critical elements such as pH and organic matter) is in the soil profile? The soil test is a key component needed for developing the nutrient rate recommendation.
- *Crop Sequence:* Did the crop that grew in the field last year (and in many cases two or more years ago) fix nitrogen for use in the following years? Has long-term no-till increased organic matter? Did the end-of-season stalk test show a nutrient deficiency? These factors also need to be factored into your plan.
- *Estimated Yield:* Factors that affect yield are numerous and complex. A field's soils, drainage, insect, weed and disease pressure, rotation and many other factors differentiate one field from another. This is why using historic yields is important in developing yield estimates for next year. Accurate yield estimates can dramatically improve nutrient use efficiency.

- *Sources and Forms:* The sources and forms of available nutrients can vary from farm-to-farm and even field-to-field. For instance, manure fertility analysis, storage practices and other factors will need to be included in a nutrient management plan. Manure nutrient tests/ analysis are one way to determine the fertility of it. Nitrogen fixed from a previous year's legume crop and residual affects of manure also effects rate recommendations. Many other nutrient sources should also be factored into this plan.
- *Sensitive Areas:* What's out of the ordinary about a field's plan? Is it irrigated? Next to a stream or lake? Especially sandy in one area? Steep slope or low area? Manure applied in one area for generations due to proximity of dairy barn? Extremely productive—or unproductive—in a portion of the field? Are there buffers that protect streams, drainage ditches, wellheads, and other water collection points? How far away are the neighbours? What's the general wind direction? This is the place to note these and other special conditions that need to be considered.
- *Recommended rates:* Here's the place where science, technology, and art meet. Given everything you've noted, what is the optimum rate of N, P, K, lime and any other nutrients? While science tells us that a crop has changing nutrient requirements during the growing season, a combination of technology and farmer's management skills assure optimum nutrient availability at all stages of growth. No-till corn generally requires starter fertilizer to give the seedling a healthy start.
- *Recommended timing:* When does the soil temperature drop below 50 degrees? Will a N stabilizer be used? What's the tillage practice? Strip-till corn and no-till often require different timing approaches than seed planted into a field that's been tilled once with a field cultivator. Will a starter fertilizer be used to give the seedling a healthy start? How many acres can be covered with available labour (custom or hired) and equipment? Does manure application in a farm depend on a custom applicator's schedule? What agreements have been worked out with neighbours for manure use on their fields? Is a neighbour hosting a special event? All these factors and more will likely figure into the recommended timing.
- *Recommended methods:* Surface or injected? While injection is clearly preferred, there may be situations where injection is not feasible (*i.e.* pasture, grassland). Slope, rainfall patterns, soil type, crop rotation and many other factors determine which method is best for optimizing nutrient efficiency in farms. The combination that's right in one field may differ in another field even with the same crop.
- *Annual review and update:* Even the best managers are forced to

deviate from their plans. What rate was actually applied? Where? Using which method? Did an unusually mild winter or wet spring reduce soil nitrate? Did a dry summer, disease, or some other unusual factor increase nutrient carryover? These and other factors should be noted. It's easier to make notes throughout the year than to remember back six to 10 months.

When such a plan is designed for animal feeding operations (AFO), it may be termed a "manure management plan." In the United States, some regulatory agencies recommend or require that farms implement these plans in order to prevent water pollution.

The U.S. Natural Resources Conservation Service (NRCS) has published guidance documents on preparing a comprehensive nutrient management plan (CNMP) for AFOs.

The International Plant Nutrition Institute has published a 4R plant nutrition manual for improving the management of plant nutrition. The manual outlines the scientific principles behind each of the four R's or "rights" and discusses the adoption of 4R practices on the farm, approaches to nutrient management planning, and measurement of sustainability performance.

CROP NUTRITION AND FERTILIZER REQUIREMENTS

ESSENTIAL PLANT NUTRIENTS

Proper nutrition is essential for satisfactory crop growth and production. The use of soil tests can help to determine the status of plant available nutrients to develop fertilizer recommendations to achieve optimum crop production. The profit potential for farmers depends on producing enough crop per acre to keep production costs below the selling price. Efficient application of the correct types and amounts of fertilizers for the supply of the nutrients is an important part of achieving profitable yields.

There are at least 16 elements known to be essential for plant growth. Carbon (C), hydrogen (H), and oxygen (O) are derived from carbon dioxide (CO_2) and water (H_2O).

Nitrogen (N), phosphorus (P), potassium (K), sulphur (S), calcium (Ca), magnesium (Mg), boron (B), chlorine(Cl), copper (Cu), iron (Fe), manganese (Mn), molybdenum (Mo) and zinc (Zn) are normally derived from the soil in the form of inorganic salts. Ninety-four to 99.5 per cent of fresh plant material is made up of carbon, hydrogen and oxygen. The other nutrients make up the remaining 0.5 to 6.0 per cent.

Macronutrients refer to those elements that are used in relatively large amounts, whereas micronutrients refer to those elements that are required in relatively small amounts.

Table. Essential Plant Nutrients

Supplied from air and water	Supplied from soil and fertilizer sources Macronutrients	Micronutrients
Carbon (C)	Nitrogen (N)	Zinc (Z)
Hydrogen (H)	Phosphorous (P)	Copper (Cu)
Oxygen (O)	Potassium (K)	Iron (Fe)
	Sulphur (S)	Maganese (M)
	Calcium (Ca)	Boron (B)
	Magnesium (Mg)	Chlorine (Cl)
		Molybdenum (Mo)
		Cobalt (Co)

All 13 elements must be present in the soil for plant use, in varying degrees of availability, to ensure both the immediate and long term needs of the crop.

Some of the commonly used terms to describe levels of nutrient elements in plants include:

- *Deficient:* When an essential element is at a low concentration that severely limits yield and produces more or less distinct deficiency symptoms. Extreme deficiencies will lead to death of the plant.
- *Insufficient:* When the level of an essential plant nutrient is below that required for optimum yields or when there is an imbalance with another nutrient. Symptoms of this condition are seldom evident.
- *Sufficient:* When the concentration of an essential nutrient is present in adequate amounts for optimum crop growth.
- *Excessive:* When the concentration of an essential plant nutrient is sufficiently high to result in a corresponding shortage of another nutrient.
- *Toxic:* When the concentration of either essential or other elements is sufficiently high to reduce plant growth severely. Severe toxicity will result in death of plants.

TYPES OF FERTILIZERS

Fertilizer refers to any compound that contains one or more chemical elements, organic or inorganic, natural or synthetic, that is placed on or incorporated into the soil or applied to directly onto plants to achieve normal growth. The main supply sources of plant nutrients include organic manures, plant residues, biological nitrogen fixation and commercial inorganic fertilizers.

The type of fertilizers that are most commonly used for crop production in Alberta are chemical fertilizers. Chemical fertilizers refer to commercially manufactured products containing a substantial amount of one or more plant nutrients.

The chemical fertilizers can be broadly classified into: nitrogen, phosphorus, and potassium fertilizers. A straight fertilizer contains only one of the nutrients. A compound fertilizer contains two or more nutrients. A complex fertilizer that is formed by mixing ingredients that react chemically, as opposed to a mechanical mixture of two or more fertilizers. A low analysis fertilizer product contains a low percentage of nutrients, usually 30 per cent or less and a high analysis fertilizer contains more than 30 per cent.

Types of popular chemical fertilizer products used in Western Canada and, average nutrient contents

Table. Percentage of Nutrients in Selected Fertilizers

Analysis Sulphur (S)	**Nitrogen (N)**		**Phosphate (P_2O_5)**		**Potash (K_2O)**
Nitrogen Fertilizers					
Urea	46-0-0	46	0	0	0
Ammonium Nitrate-(Granular)	34-0-0	34 0	0	0	
Ammonium Sulphate-Urea	34-0-0	34	0	0	11
Ammonium Sulphate	21-0-0	21	0	0	24
Anhydrous Ammonia (gas)	82-0-0	82	0	0	0
Urea-Ammonium Nitrate Solution	28-0-0	28	0	0	0
Phosphate Fertilizers					
Mono-Ammonium Phosphate	12-51-0	12	51	0	1.5
Mono-Ammonium Phosphate	11-55-0	11	55	0	0
Ammonium Polyphosphate Solution	10-34-0	10	34	0	0
Nitrogen Phosphates					
Ammonium Phosphate Sulphate	16-20-0	16	20	0	14
Ammonium Nitrate Phosphate	23-23-0	23	23	0	0
Ammonium Nitrate Phosphate	27-14-0	27	14	0	0
Urea Ammonium Phosphate	27-27-0	27	27	0	0
Urea Ammonium Phosphate	34-17-0	34	17	0	0
Potash Fertilizers					
Potash Chloride	0-0-60	0	0	60	0
Potassium sulphate	0-0-52-12	0	0	52	12
Sulphur Fertilizers					
Ammonium Sulphate	20-0-0-(24)	20	0	0	24
Gypsum (agricultural)	0-0-0	0	0	0	17
Elemental Sulphur*	0-0-0	0	0	0	90-99
Ammonium Thiosulphate Solution	12-0-0-(26)	12	0	0	26

- Elemental sulphur is suitable for application on perennial forage but is less suitable for annual crops because of slow-release characteristics. If used on annual crops it should be applied in the fallow year or in the fall to allow conversion to the sulphate form. All fertilizers must show the guaranteed nutrient analysis on the label. This states the content of three main nutrients: nitrogen, phosphate (P_2O_5) and potash (K_2O)(a form of potassium). It is shown by a series of three numbers. For example, if the numbers 10-10-10 appear on a 30 kg bag of fertilizer it means that the bag contains 10 per cent of each raw material (3 kg of nitrogen, 3 kg of phosphate and 3 kg of potash). Agriculture and Agri-Food Canada inspectors make regular checks of fertilizer facilities to ensure that all requirements are met. In addition, the sale of all fertilizer materials in Canada is regulated by the Agriculture Inspection Directorate of Agriculture and Agri-Food Canada under the authority of the Fertilizer Act.

SOIL NUTRIENT CONTENT AND SOIL TESTING

Nitrogen and phosphorus are the most commonly deficient nutrients in Canadian prairie soils. Potassium and sulphur deficiencies occur in particular areas and soil types. Calcium and magnesium are contained in lime which is plentiful in most prairie soils and therefore deficiency problems are rare. Research has found micronutrient deficiency problems are not common on the prairies. However, specific soil conditions have been identified where inadequate levels of micronutrients occur. In central Alberta, on the Black, Gray-Black transition soils and organic soils, Copper (Cu) deficiency problems and significant responses to Cu fertilizer have been observed.

Soil sampling and testing can give an excellent inventory of plant available nutrients and other soil chemical factors important for crop production. This inventory is a basis for recommending additional nutrients for crop production on an individual field basis.

Soil nutrient levels vary from year to year, and frequently will vary within fields, even on fields that seem to be uniform. It is therefore necessary to follow certain recommended steps for soil sampling and testing to develop a sound ongoing soil fertility management programme.

An understanding of general nutrient status can be obtained for a field if soil tests are conducted. Nitrogen soil testing is recommended annually as the available nitrogen can change considerably from year to year. Changes are dependent upon environmental conditions such as rainfall and temperature patterns during a growing season, type and yield of crop harvested, date of harvest, fall tillage, amount of fertilizer applied to the previous crop. Potassium and phosphorus levels do not change substantially in a soil over a period of several years. Therefore, sampling for phosphorus and potassium may be conducted every 2 or 3 years, or when changing crop type. Sampling for sulphur should be done annually, unless a previous test indicates that the available levels to 60 cm (24 in) are in excess of 55 kg/ha (50 lb/ac). In the latter case, a test for sulphur every 2 to 3 years is adequate.

While it is recognized that soil testing is not an infallible guide to crop production and that other factors also come into play, soil tests help to reduce the guesswork in fertilizer practices. In the past, Alberta farmers that have soil tested, have generally applied significantly more fertilizer than farmers that did not soil test. Past observations suggest that Alberta farmers that do not soil test in many instances may be applying less nitrogen fertilizer than needed to obtain the most economical yield.

Poor soil sampling technique is a major problem which causes variation in fertilizer recommendations. Soil testing is only as good as the quality of the soil samples. Therefore, good soil sampling procedures must include:

1. Determine where and how to soil sample each field.
2. Use proper equipment and supplies.
3. Sample at the proper time of the year.
4. Obtain samples from the correct soil depths.
5. Handle soil samples properly.

Being familiar with the proper soil sampling procedures is important whether you are doing the sampling yourself or it is being done by a custom operator.

Determine where and how to soil sample each field Soil sampling should be done on an individual field basis. Samples from different fields should not be mixed. Begin by evaluating each field to determine representative areas. Sample hilly fields with knolls, slopes, or depressions, separatedly from mid slope positions to ensure any potential sulphur or other nutrient problem is detected.

Major areas within fields having distinctly different soil properties such as texture should be sampled and fertilized as separate fields because of different nutrient requirements.

Problem areas such as saline spots, poorly drained depressions, and eroded knolls should not be sampled unless they represent a significant portion of the field.

If they do, separate samples should be obtained. In addition, other abnormal areas such as old manure piles, burnpiles, haystacks, corrals, fence rows or farmstead sites should be avoided. At least 15 to 20 sampling sites are required for each field to give a good representative sample. Samples taken from only four or five sites in a field are generally not representative and often result in incorrect fertilizer recommendations.

There are four basic methods for taking soil samples:

1. *Benchmark Soil Sampling* - This method involves selecting uniquely different areas within a field and sampling each area separately. Unique areas are selected based on soil types, topography and crop growth. Once sites are selected, the producer will take soil samples from each specific area each year to use as a guide of fertilizing all similar areas within the field. This method is rapidly gaining popularity in Alberta, particularly with farmers that are adopting precision farming techniques. This is the current recommended method which has the best chance to maximize you're economic yield.
2. *Grid Soil Sampling* - With this method, a field is sampled in an organized grid pattern. Soil sample frequency may range from taking one sample in 0.5 acre units of the field to one sample for each 5.0 acre units of the field. The smaller the soil sampling unit the greater

the accuracy of the sample. The advantage of this method is that a field map that can be prepared for each nutrient and be used for variable rate fertilization and precision farming. The cost of taking the soil samples and the soil analysis is very high and therefore is not economical for many producers.

3. *Topographic Soil Sampling* - With this method, a producer selects the separate soil sampling sites based on topography. A set of soil samples is taken from each uniquely different topographic area within a field.
4. *Random Soil Sampling* - This involves taking soil samples in a random pattern across a field, generally avoiding unusual or problem soil areas within a field. Generally the field should not be more than 80 acres in size and has been cropped uniformly in the past. Normally, 15 to 20 sites must be sampled to obtain a representative soil sample of the field. This is the most common method of sampling presently used in Alberta.

Use proper equipment and supplies

A soil sampling probe is best for taking samples to the 60 cm (24 in) sampling depth. Use clean labelled plastic pails for collecting samples. Metal pails should not be used if micronutrient testing is to be done. Soil sample augers can also be used but it can be difficult to accurately separate soil samples into 0-15, 15-30 and 30-60 cm depths.

Tools may be borrowed from a fertilizer dealer, your local Alberta Agriculture district office, or may be purchased from some soil testing laboratories or fertilizer dealers. Information sheets, soil sample cartons, and shipping boxes are available from soil testing laboratories.

Sample at the proper time of the year

Ideally, samples should be taken just prior to seeding. However, from a practical standpoint, this is difficult because little time is left to plan a fertilizer programme and purchase fertilizer in time for seeding. The best alternative is to obtain soil samples taken in the fall once soil microbial activity has declined. The proper sample time in the fall is after the soil surface temperature drops to less than 7°C. At this temperature soil processes such as mineralization (breakdown of soil organic matter into plant available nutrients) that cause changes in soil nutrients proceed quite slowly and therefore changes in plant available nutrient levels are normally not great. Generally, it is safe to soil sample in most areas of Alberta after the beginning of October. By sampling in the fall, there is sufficient time to properly process samples, provide test results and recommendations and develop a fertilizer programme for that fall or the next spring.

Obtain soil samples from the correct depths

Many soil testing labs suggest that a 0 to 30 cm (0 to 12 in) depth sample is adequate for developing fertilizer recommendations. However, for ideal plant nutrient evaluation it is suggested that samples be taken from the 0 to 15 cm (0 to 6 in) and 15 to 30 cm (6 to 12 in), separately and also take samples from the 30 to 60 cm (12 to 24 in) depths. Keep samples from each depth in a separate container.

Phosphorus and potassium recommendations are based on a measure of the amounts of the available forms of each of these nutrients contained in the 0 to 15 cm (0 to 6 in) depth sample. Generally, most of the plant available P in soil is confined to the plow layer as P is very immobile. Nitrate-nitrogen (NO_3-N) and sulphate-sulphur (SO_4-S) are both mobile nutrients may be found in significant amounts in the 30 to 60 cm depth. Therefore, N and S fertilizer recommendations based on a 0 to 15 cm (0 to 6 in) depth sample usually suggest a higher than necessary fertilizer application rate. For this reason, recommendations regarding nitrogen and sulphur are based on a measure of the amount of nitrogen and sulphur from depth (0-60 cm) soil samples.

Handle soil samples properly

As mentioned, soil from each depth should be placed in separate containers. Immediately after the samples have been taken:

1. Mix the soil in each container thoroughly in order to obtain a homogeneous mixture.
2. Remove the soil and spread on a piece of clean paper.
3. Allow the soil to completely air dry at a temperature of not more than 30°C. Do not dry in an oven or at a high temperature since this can change the levels of some nutrients.
4. Care should be taken to avoid contamination of the samples with fertilizer materials such as commercial fertilizers, manure, salt, water and dust. Samples should not be dried on old fertilizer or feedbags or areas where fertilizers have been handled.
5. A fan may be used to ensure constant air flow over samples to enhance drying.

Once the samples are thoroughly dry, fill the soil sample cartons. Label each carton with correct field number and sample depth. Complete an information sheet on cropping and fertilizer history and mail to a reputable soil testing laboratory. The services of soil testing and fertilizer recommendations are not free. Consult your Alberta Agriculture crop specialist for details of private soil testing laboratories that are available.

Fertilizer recommendations

Fertilizer recommendations are based on the results of the soil test analyses and on the nutrient requirement of the crop to be grown. Recommendations on time and method of fertilizer application are also included. Each soil testing lab has its own philosophy for making fertilizer recommendations. Two examples are:

1. Recommendations which indicate the nutrient requirements and yield potentials for optimum economic production based one or more moisture conditions of the field.
2. "Target Yield Recommendations" which indicate the nutrient requirements for a range of various lower and higher yield potentials under the same moisture conditions. With this information the producers have the flexibility of selecting a fertilizer application rate or target yield that best suits their individual situation.

Producers must keep in mind that optimum yields of high quality spring wheat can not be obtained without adequate fertilization if the crop to be grown on soils deficient in essential elements. Fertilization, however, will neither increase yield or quality of wheat if other management inputs and cultural practices are not optimal, nor will it increase yield if the added nutrients are not required. Therefore, the most successful fertilizer programme will be based on a knowledge of soil nutrient status combined with optimum crop and fertilizer management practices.

Plant Tissue Analysis

Plant tissue analysis measures nutrient levels in the plant during their growth. The supply of available nutrients is reflected in the nutrient content of the crop.

Therefore use of plant tissue analysis allows a producer to evaluate the effectiveness of fertilizer recommendations from a soil testing service. Producers who do not soil test can still use routine plant tissue analysis to evaluate their fertilizer management programme to determine whether they used the correct kinds and amounts of nutrient.

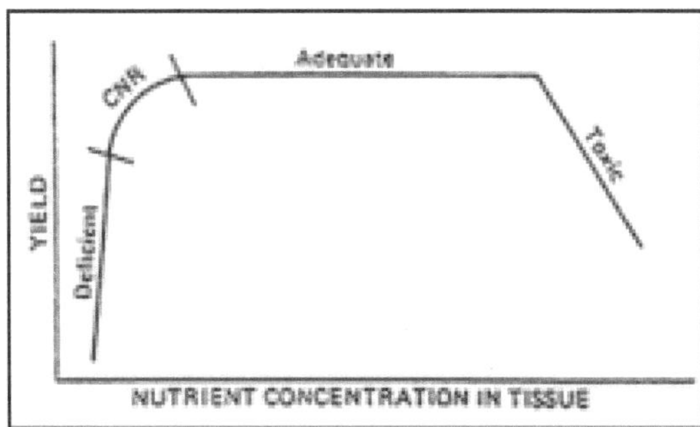

Fig. Hidden Hunger is a term used to describe a plant that shows no obvious symptoms, yet the nutrient content is not sufficient to give the top profitable yield. fertilization with the "sure" rate rather than the bare economic optimum for an average year helps to obtain the top profitable yield.

Plant tissue analysis can be used to diagnose crop nutrition production problems. For example, at times, a plant growth problem occurs and can not be explained. Soil, climatic and other environmental conditions seem favourable, essential plant nutrients were supplied, and other sound management practices were followed, plant tissue analysis can indicate if nutrient levels may be associated with the problem.

For reliable results and useful interpretation, producers must follow proper procedures for plant sampling and sample handling. The sample must represent the crop it was taken from. Management decisions based on plant tissue analysis results reflect the quality of the sample provided.

In sampling wheat plants, parts to sample are the entire above-ground plant. Ideally, tissue samples must be taken when the plant is at Zadok's growth stage 30. Growth stage 30 is when the leaf sheath of wheat has begun to lengthen and are strongly erected, but before the first node of the stem is visible at the base of the plant. If the sample is taken prior to growth stage 30, nitrogen is more concentrated in the leaf, so the prescription would call for too little nitrogen to be applied. If taken after growth stage 30, rapid growth has begun and nitrogen content will be diluted by increases in dry matter resulting in excessive nitrogen fertilizer estimates.

An adequate sample must contain 20 to 50 individual plants. Samples should be taken in the morning (or on cool or cloudy days) because heat and moisture stress often occurs during the mid-day and mid afternoons, on hot sunny days or immediately following a rain. Plant samples should be cut with a clean, sharp, rust free knife, blade or scissors. Clean plastic containers should be used for sample collection. In subsequent handling, clean brown paper bag can be used.

After sampling:

1. Remove loose soil particles with a clean, dry cloth or brush.
2. Do not wash the sample.
3. Lay out the sample on clean paper until completely air dried at normal room temperature (25 to 35°C).
4. Do not oven dry samples.

Completely dried samples can then be placed in a plant tissue sample bag or clean brown paper bag. Information Sheets from a testing lab should be completed for each sample as this information helps when interpreting the results. The sample bag and the corresponding information sheet should each be carefully labelled with the same identity so that samples and sheets can be matched in the laboratory.

The laboratory will determine the levels of each nutrient in the plant sample, and will indicate if each nutrient level is excessive, adequate, marginal or below a critical level. The critical concentration of a nutrient is the point where crop growth may be 10 per cent or less than the maximum. This system uses a previously established set of standards for nutrients in a specific plant part, sampled at a particular growing stage and then compared those normals to the findings in the sample. The principle of this system lies in the relationship between nutrient concentration and crop yield.

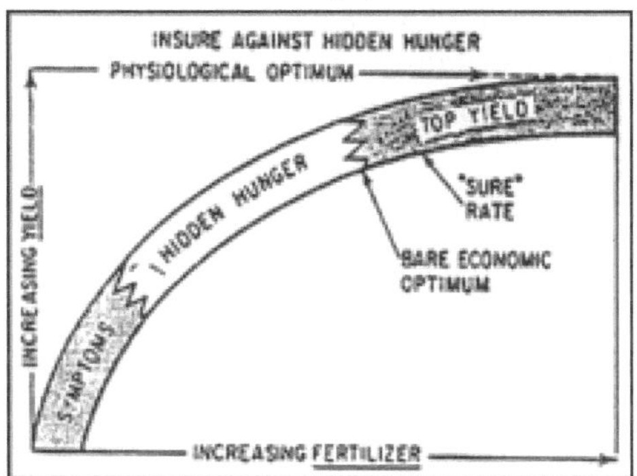

Fig. Relationship between nutrient concentration in plant tissue and top yield, showing the proposed critical nutrient range. (CNR)

The most meaningful definition of critical nutrient concentration for efficient growers is the level of a nutrient below which crop yield, quality, or performance is unsatisfactory. In spring wheat, 2.0-3.0 per cent N, 0.26-0.5 per cent P, and 1.5-3.0 per cent K in the whole plant prior to filling are considered sufficient.

Table. Plant Tissue Analysis Interpretative Critera Crop Cereals Plant Part/Growth Stage Whole Plant Prior to Filling

Nutrient		Low	Marginal	Sufficient	High	Excess
Spring		1.5	1.5 - 2.0	2.0 - 3.0	3.0 - 4.0	4.0
Nitrogen (N) %	Winter	1.25	1.25 - 1.75	1.75 - 3.0	3.0 - 4.0	4.0
Phosphorous (P) %		0.15	0.15 - 0.25	0.26 - 0.5	0.5 - 0.8	0.8
Potassium (K) %		1.0	1.0 - 1.5	1.5 - 3.0	3.0 - 5.0	5.0
Sulphur (S) %		0.1	0.1 - 0.15	0.15 - 0.40	0.40 - 0.8	0.8
Calcium (Ca) %	Other	0.10	0.10 - 0.2	0.2 - 1.0	1.0 - 1.5	1.5
	Barley	0.20	0.2 - 0.3	0.3 - 2.0	2.0 - 2.5	2.5
Magnesium (Mg) %		0.1	0.1 - 0.15	0.15 - 0.50	0.5 - 1.0	1.0
Zinc (Zn) ppm		10	10 - 15	15 - 70	70 - 150	150
Copper (Cu) ppm	Barley	2.3	2.3 - 3.7	3.7 - 25	25 - 50	50
Wheat		3.0	3.0 - 4.5	4.5 - 25	25 - 50	50
	Oats	1.7	1.7 - 2.5	2.5 - 25	25 - 50	50
Iron (Fe) ppm		15	15 - 20	20 - 250	250 - 500	500
Manganese (Mn) ppm		10	10 - 15	15 - 100	100 - 250	250
Boron (B) ppm		3	3 - 5	5 - 25	25 - 75	75
Molybdenum (Mo) ppm		0.01	.01 -.02	.03 - 5	5 - 10	10

The CNC (critical nutrient concentration) system is the commonly used method in interpreting plant tissue analysis in the prairie provinces as well as other parts of the world. However there are some limitations associated with this system. Some of the limitations include:

Low Sensitivity: The diagnosis requires that the nutrient concentration of plant tissue be compared with previously set values, usually in the form of permissible ranges of variation for each element at a particular stage of growth. These ranges are often very wide, and are therefore not very sensitive.

Inconvenience: It is not always convenient to take samples at a particular physiological stage of the plant's life.

Plant Growth Stage: Inadequacy in dealing with the variations in nutrient concentration with the stage of the plant growth and which plant parts that are sampled.

Multiple Growth Stress: Inability to establish the relative importance of the limiting factors in terms of yield when more than one nutrient being diagnosed is deficient. If other stresses such as drought, salinity, diseases or insects are causing plant growth problems, these can affect plant nutrient concentration in the tissue.

ZINC AND MANGANESE MANAGEMENT IN CALCAREOUS SOILS

Soil pH is the most important factor regulating Zn and MN supply in calcareous soils. At alkaline pH values, very low levels of soluble Zn are found, and therefore only a negligible amount can be in the form of exchangeable Zn^{2+}, which is available to plants. Zinc and MN deficiencies are clearly pH-dependent, and both Zn and Mg concentration in solution decreases 100-fold for each unit increase in pH. At high pH levels precipitates are formed, such as Zn hydroxides [$Zn(OH)_2$ and $CaZn(OH)_4$] and Zn carbonate.

Soils that are slightly alkaline may not necessarily be deficient in Zn or MN In addition, Zn and MN can be chelated by natural organic compounds in the soil, a process that aids the movement of these nutrients to the plant root. On highly alkaline soils, however, Zn and MN deficiencies are not uncommon. The most common inorganic Zn and MN fertilizers are the sulfates ($ZnSO_4$, $MnSO_4$) and their oxide forms (ZnO, MnO). Broadcast application of these compounds to correct Zn or MN deficiencies in calcareous soils is not effective, since the alkaline pH renders the Zn and MN unavailable almost immediately.

Zinc is also available in chelated forms, including Zn-EDTA and Zn-EDDHA. Chelated Zn when applied to calcareous soils remains soluble and available to plants considerably longer than the inorganic forms. However, soil application of chelated Zn is rarely economical. Manganese chelates have limited effectiveness in calcareous soils and are not normally used.

The least expensive way to apply Zn and MN is through foliar sprays. In addition to the forms listed above, a number of other Zn and MN formulations are available for foliar spraying, including nitrates and organically chelated forms using lignin sulfonate, glucoheptonate, or alpha-keto acids. Preliminary research data indicate little difference in magnitude of foliar uptake, regardless of the form of carrier or chelate applied. Foliar applications of low rates of MN or Zn (*e.g.*, 0.6-1.2 kg ha^{-1}) are not adequate to correct moderate to severe deficiencies often found in soils with high pH values. Foliar application of Zn has to be repeated several times per year, due to the limited translocation from older to new leaves.

COPPER, BORON AND MOLYBDENUM MANAGEMENT IN CALCAREOUS SOILS

Copper: Copper solubility is pH-dependent and it decreases with increasing pH.

However, the appreciable changes in solubility occur in the pH range below 5, and above pH 5 the solubility does not change very much. Copper deficiencies are generally found in leached sandy soils. The most common source of Cu for soil and foliar application is copper sulfate ($CuSO_4 \bullet 5H_2O$) and also Cu chelates. In orchards, the crop Cu requirement is normally satisfied through foliar sprays of Cu fungicides.

Boron: Soil pH affects B availability more by sorption reactions than by formation of less soluble compounds. Availability of B is highest in the pH range of 5.5-7.5. Boron is sorbed to Fe and Al oxides in soils and is lowest in the pH range of 6-9.

There is an interaction between B availability and the presence of Ca ions. High levels of Ca at high pH reduce the uptake of B. This may explain the fact that high B levels in calcareous soils, considered as toxic in other conditions, do not produce B toxicity in crops.

Responses to boron application in sandy calcareous soils have been reported for sugar beet and bitter orange. In deficient soils, B is most applied in the form of sodium borates ($Na_2B_4O_7 \bullet xH_2O$). Excess and toxicity of B usually occur in soils with high levels of B or due to additions of B in irrigation water.

Molybdenum: Mo deficiencies are not known in calcareous soils, as MO availability increases with pH. Molybdenum toxicity to plants has not been reported. However, high MO levels can accumulate in alkaline soils, which may be toxic to livestock this condition is known as *molybdenosis*.

NUTRIENTS AVAILABILITY IN THE RHIZOSPHERE

Mobilization of P and micronutrients in the rhizosphere of plants grown on calcareous soils can be brought by different mechanisms: (a) Root-induced decrease in pH as a consequence of preferential cation uptake; (b) Exudation of organic acids by the roots; (c) Release by roots of photosynthates as substrate for rhizosphere microorganisms, which in turn affect pH, redox potential, and chelator concentrations (*e.g.* siderophores) in the rhizosphere.

Roots can affect P and micronutrients availability by inducing pH changes in the rhizosphere. The pH at the root surface may differ from the bulk soil pH by up to two units. Such differences in pH are caused by differential rates of uptake of cations and anions, which are associated with OH^- and H^+ effluxes.

The form of N nutrition plays an important role. In plants supplied with NO_3^-, anions uptake exceeds cation uptake and OH^- or HCO_3^- is released from the roots. This causes the rhizosphere pH to become more alkaline. On the other hand, plants supplied with NH_4^+ take up more cations than anions, release H^+ into the soil, thus acidifying the rhizosphere and increasing P and micronutrient availability. N- anionic nutrition (*i.e.* supply of NO_3^-) was found to cause Fe chlorosis in calcareous soils by inducing high pH levels in the free space of root and leaf tissue and thereby restrict the uptake of Fe into the cell. Shifting to N-cationic nutrition (*i.e.* application of NH_4^+) induce Fe chlorotic maize leaves to re-green without application of Fe.

Plants roots also exudates low molecular weight compounds to the rhizosphere, such as sugars, organic acids, amino acids and phenolics. Organic acids, such as citric and malic acid, are of general importance in mobilizing P, Fe MN and Zn in the rhizosphere, and can increase their uptake rates and contents in the plants. In response to Fe deficiency, increased rates of root exudation of phenolics and amino acids (phytosiderophores) play a particularly significant role.

Enhanced exudation of organic acids by tomato roots was measured under P deficiency conditions. Exudation of organic acids can cause localized acidification of the rhizosphere; this is specially observed in species with clustered roots (*proteoid* roots) such as white lupin. Citric acid was the dominant compound in proteoid roots of white lupin, and increased Fe, MN and Zn mobilization from calcareous soils. The increased availability of these nutrients by exuded organic acids stems either from the lower pH of the rhizosphere, or the chelation of Fe, MN and Zn, and by lowering the Ca^{2+}concentration by chelation and formation of sparingly soluble salts such as Ca citrate.

INTEGRATING COVER CROPS INTO GRAPEVINE PEST AND NUTRITION MANAGEMENT

COVER CROPS

Cover crops are being studied because they may have substantial effects on several components of agroecosystems. In various locations cover crops have been used to enhance the natural control of arthropod pests, to suppress weeds, and as an alternative to synthetic nitrogen fertilizer. These uses of cover crops are usually examined independently, but there is reason to consider the interactions among them as well.

In fact, cover crops can impact agroecosystems through a number of pathways, both direct and indirect. If, for example, cover crops affect the nutritional quality (and water status) of crop plants, then we should expect these effects to also impact the performance of herbivores feeding on the crop plant, since plant nutrient and water status can significantly affect population dynamics of herbivores. Moreover, these indirect effects may increase incrementally through time, as the cover crop system undergoes a transition phase. A long-term, comprehensive assessment of the role of cover crops in crop production is necessary if we are to understand and successfully utilize them for arthropod, weed, and crop-nutrition management.

The research reported in this article assessed the value of cover crops in vineyard pest and nutrition management. Four experiments helped us determine the effect of various vineyard floor management systems on: 1) two species of leafhoppers, the variegated grape leafhopper (*Erythroneura variabilis* Beamer), and the western grape leafhopper (*Erythroneura elegantula* Osborn), which are major pests of grapevines in the San Joaquin Valley: 2) weed suppression and the need for soil-applied herbicides; 3) vine-nutrient status and the need for synthetic nitrogen fertilizer; and 4) vine growth, and grape yield and quality. We also developed a partial budget for each system.

EXPERIMENTAL DESIGN AND CULTURAL METHODS

Fowler and Earlimart sites

Two experiments were initiated in 1992 comparing three vineyard floor management systems: 1) a 'conventional' clean-cultivated system with chemical weed suppression in row berms, 2) a fall-planted, oat-vetch cover crop mix which was mowed and placed on row berms for weed suppression without the use of soil-applied herbicides, and 3) a fall-planted, oat-vetch cover that was mowed and placed in row middles, with complete chemical weed control in row berms.

These systems were compared at two locations: a 60-acre 'Thompson Seedless' vineyard located about eight miles west of Fowler (Fresno County), and a 33-acre 'Thompson Seedless' vineyard located about ten miles east of Earlimart (Tulare County). In both cover crop systems, the cover was mowed twice and then allowed to re-seed before soil incorporation, which typically occurred in early July at the Fowler site, and October at the Earlimart site. All systems were clean-cultivated for the remainder of the season at the Fowler site (a raisin grape vineyard), while a ground cover was maintained during the summer by periodic mowing of resident vegetation in all plots at the Earlimart site (a table grape vineyard).

Insecticide and miticide use differed between the Fowler and Earlimart sites. At the Fowler site, a low rate of Omite (two pounds per acre) was applied on 14 June, 1993 to control an outbreak of Pacific spider mite (*Tetranychus pacificus* McGregor). No other miticides or insecticides were applied for leafhopper and mite control at any other time during the experiment. At the Earlimart site, Pyrenone was applied in late May 1992 and 1993, and Lannate was applied in mid-August 1993, despite low numbers of leafhoppers. In 1994, the vineyard operator withheld insecticide applications for leafhopper control. Miticides were not used at any time during the experiments at the Earlimart site, where spider mites generally occurred in low densities.

Soil amendments also differed between the two sites. All plots at the Earlimart site received a yearly application of four tons per acre of steer manure in the fall, and two tons per acre of gypsum in June. At the Fowler site, all plots received 75 pounds of nitrogen per acre in June 1992, and four tons per acre of steer manure in October 1992. During the 1993 season, we added fertilized and unfertilized subplots to each treatment. Fertilized subplots received 75 pounds nitrogen per acre in June, and unfertilized plots did not receive any nitrogen fertilizer. The same design was maintained during the 1994 season, but nitrogen fertilizer was not added as vine nitrogen levels were adequate in all treatments.

Kearney and Madera sites

In 1993 we initiated two additional experiments using a Merced rye and vetch cover crop mix: one in a two-acre 'Thompson Seedless' vineyard located at the University of California Kearney Agricultural Center (Parlier, Calif.), and the other in a 53-acre commercial 'Thompson Seedless' vineyard located near Madera, Calif. The Kearney experiment compared a clean-cultivated system with a system that used the cover crop as dry mulch for partial weed suppression on row berms. The Madera experiment compared a clean-cultivated system with two cover crop systems: one in which the cover was mowed and incorporated as green manure in early April, and another in which the cover crop was mowed once (also in early April) and then allowed to reseed before incorporation in early July.

Insecticides and miticides were not used for leafhopper or mite control in the Kearney and Madera sites. Fertilizer was not added at the Kearney site, since the irrigation water at the UC Kearney Agricultural Center normally supplies vines with approximately 20 pounds of nitrogen per acre. All plots in the Madera vineyard received 20 pounds of nitrogen per acre in June 1993, but fertilizer was not added in 1994.

Sampling

We used several sampling methods (direct observations, sticky traps, D-Vac, sweep net, pitfall traps and vine shakes) to estimate the abundance of leafhoppers and their natural enemies (spiders and other generalist predators, and egg parasitoids) on vines and on the cover crops. In general, sampling of cover crops was conducted from March until plow down, while grape vines were sampled from May through September or mid-October. Controlled shoot-cage experiments were also conducted in 1993 at the Madera site, and in 1994 at the Fowler, Kearney, and Madera sites. The purpose of these cage experiments was to determine the impact of each treatment on leafhopper reproduction and survivorship in the absence of the actions of their natural enemies. We also conducted trace-element marking experiments in the Madera vineyard during the 1993 and 1994 seasons, to further determine the impact of cover crops on leafhoppers and spiders.

The effect of berm mulching with cover crop biomass on weed suppression was assessed at the Earlimart, Fowler and Kearney vineyards. Prior to each mowing and in the middle of summer, we determined weed species frequency, per cent cover and total biomass on berms, and per cent of berm area covered with mulch residues. Weed suppression was evaluated at each time by identifying and counting plant species using four 100-foot transects, and counting plants at each foot.

We used leaf petioles to determine vine nitrogen, phosphorus, and potassium status at two stages of vine growth: full bloom and beginning of berry ripening (veraison). At harvest, we estimated berry size and sugar content (all vineyards), pH and titratable acidity (Earlimart vineyard), and raisin yield and quality (Fowler, Kearney and Madera vineyards).

Results and Discussion

Leafhoppers

Our findings to date indicate that, if properly managed, winter legume/grass cover crops can reduce leafhopper abundance. Where leafhopper numbers were not very low and cover crops were properly maintained through early July, reduced infestations of leafhoppers were observed in the presence of cover crops. These reductions may be attributed in part to enhanced activity of certain groups of spiders, which are consistently found at higher densities in the presence of cover crops compared to the clean-cultivated systems.

Trace-element marking of cover crops indicated that cover crops may also influence leafhopper populations by serving as non-host vegetation which interferes with their movement patterns and perhaps other aspects of their life history (Hanna et al., unpublished data). Although cover crops may affect vine physiology (through changes in soil fertility, soil-water status, etc.), which may in turn affect leafhopper biology, the changes we have observed in vine conditions to date were not sufficient to explain the differences in leafhopper abundance observed between the cover crop and clean-cultivated treatments. Cage data indicated that leafhopper reproduction and survivorship in the absence of predation were not affected by cover crop treatment.

Vine Nutrition

The effect of cover crops on vine-nutrient status varied between years and vineyards. Cover crops produced positive effects on vine nutrient status by the second or third year if the cover crops were well-managed, but produced negative effects if the cover or vineyard was poorly managed. The positive effects were usually delayed, and were best illustrated by the results from our Fowler site where by the third year, nitrogen levels (nitrate-N in petioles) in unfertilized cover cropped vines were similar to nitrogen levels in fertilized clean-cultivated vines, independent of the type of weed management in row berms. Potassium levels were also enhanced by cover crops by the third year at the Fowler site.

We have not observed these effects in the Earlimart vineyard where the cover crop was not incorporated until fall, while weeds were allowed to grow in row middles during the summer. Under this system of ground cover management, much of the nutrient content of the cover is apparently either lost by volatilization, or used to grow the resident vegetation during the summer.

At the Kearney and Madera sites (second year), although we have not detected statistically significant changes in vine-nutrient status in the cover crop treatments compared to clean-cultivated treatments, we have observed non-significant trends towards higher levels of nitrogen and potassium in vines associated with cover crops compared to clean-cultivated vines. Based on results from the Fowler site, we expect to see higher levels of nitrogen and potassium in cover crop treatments at the Kearney and Madera sites during the third and later years, since the impact of cover crops on vine nutrient status is often delayed and may not be statistically significant during the transition phase.

Weeds

The amount of dry biomass produced by cover crops for weed suppression varied between vineyards. During late winter and early spring, the mulched berms received 1,800 to 8,726 pounds of dry biomass per acre, with a total nitrogen content of 33 to 109 pounds per acre.

Results to date from the San Joaquin Valley sites and from a north coast study indicate that with sufficient levels of biomass production, berm mulching should reduce the use of pre-emergence herbicides. The mulch, however, will not control all weeds equally. Perennial weeds such as field bindweed were not controlled.

Yield

The effects of cover crops on grape yield and operating costs depended on grape culture, and represented a trade-off in water, fertilizer, pesticide and resource use. Although significant differences in yields have not been realized in the Earlimart (third year) and Madera (second year) vineyards, berry weight and raisin yields were significantly higher by the second year at the Kearney site where cover crop biomass was used as dry mulch for weed suppression in row berms. Berry weight was also significantly greater by the third year in cover crop compared to clean-cultivated treatments at the Fowler site. Greater berry weights should have translated into greater raisin weights, but this effect could not be measured as the raisins at the Fowler site were badly damaged by early fall rain. The partial cost budget indicated that the use of cover crops (despite greater water demand) may significantly reduce operating costs if savings were realized by reducing chemical inputs for insects and mites. These savings are expected to increase if cultural methods (*e.g.*, raised beds with adjacent furrows for irrigation are used instead of flood irrigation) are modified to maintain satisfactory cover crop growth while reducing water use.

4

Fertilizers Applications and Crop Production

COMMON FERTILIZERS AND APPLICATIONS

THE CONCEPT OF FERTILIZATION

In order to enhance crop yields farmers have applied fertilization all over the world since ancient times. It is obvious that depending on natural conditions and the level of the development of agriculture in a given area, the fertilizing materials and the methods of their application have always shown considerable variation. As one of the most important aspects of agricultural technology, the practice of fertilization has undergone great developments in the past centuries. Ever since agriculture as a modern science was born, fertilization has always been one of the main focuses of scientific research.

The modern concept of fertilization (adopted from Havlin et al. 2005) can be summarized according to the following. First of all, concerning the materials used for fertilization we may say that for optimum crop growth and yield both organic and inorganic nutrient sources may be used. A system in which both sources are used could be called „Integrated Nutrient Management". As for the knowledge underlying the practice of fertilizer application, it must be clearly pointed out that efficient nutrient management requires the understanding of nutrient cycling and the transformation of the given elements. It is a very important consideration that management practices that minimize losses and maximize the amount of applied nutrient recovered by the crop will, on the one hand, reduce potential environmental impacts, and increase efficiency on the other hand.

Mineral (inorganic) fertilizers are basically chemical compounds (mineral salts of N, P, K, Ca, Mg etc. Except urea), which may come in two different forms; as solid fertilizers (prepared as granules or pulverized) orliquid (fluid) fertilizers (solutions, suspensions).

The other large groups of fertilizers is that of fertilizers of organic sources. This category includes some of the most ancient types of fertilizers, which have been used traditionally for many-many centuries. Such fertilizers are animal (farmyard) manure, green manure, crop residues and organic wastes (composts, industrial wastes, sewage sludge etc.).

THE CONCEPT OF ORGANIC FARMING

Let's go organic?

As the negative environmental impacts of industrial, intensive agricultural production are demonstrated by a huge number of research results and surveys in several regions of the world, it becomes clear that certain farming practices cannot be continued without causing environmental risks for the sustainability of the ecosystems, agricultural activities should be regulated and if required, even limited. Recently, it has become a great challenge on global, regional and national level.

It is an especially grim prospect if we consider the fact, treated also at the beginning of this textbook, that it is an all-important task for humanity to increase agricultural production for a global population growing at a dramatic pace. We simply cannot afford to destroy the natural resources of the earth as it has unfortunately been done in many parts of the world. A logical answer to this global challenge is – of course – looking for new ways. Organic agricultural practices have been developed as a way of facing that challenge.

According to the USDA (United States Department of Agriculture) National Organic Standard Board (NOSB), organic agriculture is "an ecological production management system that promotes and enhances biodiversity, biological cycles and soil biological activity". „It is based on minimal use of off-farm inputs and on management practices that restore, maintain and enhance ecological harmony".

The Concept of Organic Farming

The term *organic farming* commonly used as a synonym for *ecological farming or biological farming* defines a farming system that exclusively applies organic fertilizer substances such as farmyard manure and biological pest control without any agro-chemicals. Organic farming relies on natural biological nutrient cycles and utilizes all the achievements of expertise and knowledge from either scientific results or practical experiences.

Organic farming has numerous benefits:

- Reduces or minimizes inputs (industrial products, energy etc.)
- Ensures long-term sustainability of soil fertility
- Minimizes negative environmental impacts
- Preserves of non-agricultural habitats, has a positive impact on natural vegetation and fauna
- Preserves rural population and provides safe and healthy lifestyle for farmers and their families

International and national policies, legislation represents important parts of the wider context related to organic farming systems. The framework set up by the European Commission gives a wide range of standards (applications, supervision and also sanctions) held at national or local levels.

NUTRIENT MANAGEMENT IN ORGANIC FARMING

Organic Farming applies an internal nutrient cycle using the advantages oforganic fertilizer sources (green manure, cover crops, farmyard manure etc.), for reduced inputs and soil fertility maintenance.

Green Manuring

The term Green Manuring refers to a plant-based fertilization approach applying locally produced crops and their residues such as winter and spring cereals, legumes etc. Green manures of legumes (*e.g.* clovers/Trifolium etc.) from oversowing and undersowing are incorporated into the soil.

Cover crops

Cover crops and living mulches bring many benefits to crop production. Interest in winter annual cover crops such as winter rye and hairy vetch for ground cover and soil erosion control has been increasing in the last 30 years in several areas. According to experimental results, the integration of cover crops into a cropping system by relay cropping, overseeding, interseeding, and double cropping may serve to provide and conserve nitrogen for grain crops, reduce soil erosion, reduce weed pressure, and increase soil organic matter content. Hairy vetch has increased availability of nitrogen to succeeding crops, increased soil organic matter, improved soil structure and water infiltration, decreased water run-off, reduced surface soil temperature and water evaporation, improved weed control, and increased soil productivity.

Farmyard manure

Farmyard manure (FYM) consist of solid and fluid animal excreta (*i.e.* faeces and urine) and bedding materials.

Farmyard manure refers to the decomposed mixture of dung and urine of farm animals. On an average, the nutrient content of decomposed farmyard manure is 0.5-0.6 per cent N, 0.2-0.3 per cent P2O5 and.0.5-0.6 per cent K2O. The method of preparing farmyard manure by the farmers is often defective. Urine, which is wasted, contains about 1 per cent N and 1.35 per cent potassium. Nitrogen present in urine is mostly in the form of urea which is subjected to volatilization losses. Even during storage, nutrients are lost due to leaching and volatilization.

Most common farmyard manures:

Cattle shed wastes-dung, urine and slurry from biogas plants

Poultry Jitter, horse, sheep and goat manure

Average nutrient content of urine: water content is over 90 per cent (several compounds containing N → approximately 80 per cent urea, relatively high K and low P content)

Table. Annual manure production of different types of livestock

Type of livestock	Average manure Production
Beef cattle	10-11
Cow	9-10
Calf	3-4
Swine/fattening	1.0-1.2
Swine	0.6-0.8
Horse	5.0-5.5
Sheep	0.4-0.5

Average nutrient content of livestock manure (%)

Name	Water	Dry matter	N	P_2O_5	K_2O
		Faeces			
Cattle	80 – 85	13 – 18	0.3 – 0.6	0.2 – 0.3	0.1 – 0.2
Swine	75 – 85	13 – 20	0.5 – 0.7	0.4 – 0.6	0.3 – 0.5
Sheep	60 – 70	25 – 35	0.5 – 0.7	0.3 – 0.4	0.1 – 0.2
Horse	73 – 77	20 – 23	0.5 – 0.6	0.3 – 0.4	0.3 – 0.4
		Urine			
Cattle	90 – 93	3 – 6	0.6 – 1.0	0.10 – 0.15	1.0 – 1.5
Swine	94 – 97	2 – 3	0.5 – 0.6	0.05 – 0.15	0.8 – 1.0
Sheep	87 – 91	7 – 8	1.4 – 1.6	0.10 – 0.15	1.5 – 2.0
Horse	89 – 93	5 – 7	1.2 – 1.4	0.01 – 0.05	1.5 – 1.8

Efficiency of farmyard manure

Nutrient supplying of FYM is usually last for at least two years, depending on several factors such as soil and climatic conditions and manure characteristics.

Table. Average expected nutrient content of 10 tons of FYM in kg

	N	P_2O_5	K_2O
1st Year	18	20	40
2nd Year	12	15	20

Average efficiency (expressed in percentages), demonstrating the differences in duration depending on FYM quality and soil characteristics.

Table. Efficiency of FYM expressed in % in a sandy and sandy loam soil

Long-term effect	N			P_2O_5			K_2O		
	Poor*	medium*	good*	poor	medium	good	poor	medium	good
	*Manure quality								
1st year	50	42	42	50	50	50	50	50	50
Second year	33	42	42	35	33	38	38	33	30
Third year	17	16	16	15	17	12	12	17	20

Table. Efficiency of annually applied FYM on clayey loam and clay soils independently of manure quality

Nutrient	Efficiency expressed in percentages			
	1st year	Second year	Third year	Fourth year
Nitrogen	40	70	90	100
Phosphorus	33	66	80	100
Potassium	33	66	80	100

MINERAL FERTILIZER SOURCES

The term mineral fertilizer refers to industrial products.

The following are the fertilizers commonly used in modern agriculture in the production of various crops all over the world:

- Single nutrient fertilizers (One nutrient element)
  Macroelement fertilizers (N, P, K, Mg, Ca, S)
  - Microelement fertilizers (B, Fe, Mn, Mo, Cu, Zn etc.)
- Multielement (Mixed, complete, complex) fertilizers
  - N + P + K + Ca + Mg + S (+ microelements) - For balanced crop nutrition

FERTILIZER UREA

Urea, a white crystalline solid containing 46 per cent nitrogen, is widely used in the agricultural industry as an animal feed additive and fertilizer.

PHYSICAL FORMS OF UREA

Commercially, fertilizer urea can be purchased as prills or as a granulated material. In the past, it was usually produced by dropping liquid urea from a "prilling tower" while drying the product. The prills formed a smaller and softer substance than other materials commonly used in fertilizer blends. Today, though, considerable urea is manufactured as granules. Granules are larger, harder, and more resistant to moisture. As a result, granulated urea has become a more suitable material for fertilizer blends.

Advantages of fertilizer urea

- Urea can be applied to soil as a solid or solution or to certain crops as a foliar spray.
- Urea usage involves little or no fire or explosion hazard.
- Urea's high analysis, 46 per cent N, helps reduce handling, storage and transportation costs over other dry N forms.
- Urea manufacture releases few pollutants to the environment.
- Urea, when properly applied, results in crop yield increases equal to other forms of nitrogen.

INCORPORATE UREA FOR BEST USE

Nitrogen from urea can be lost to the atmosphere if fertilizer urea remains on the soil surface for extended periods of time during warm weather. The key to the most efficient use of urea is to incorporate it into the soil during a tillage operation. It may also be blended into the soil with irrigation water. A rainfall of as little as 0.25 inches is sufficient to blend urea into the soil to a depth at which ammonia losses will not occur.

Urea losses to air

Urea breakdown begins as soon as it is applied to the soil. If the soil is totally dry, no reaction happens. But with the enzyme urease, plus any small amount of soil moisture, urea normally hydrolizes and converts to ammonium and carbon dioxide. This can occur in 2 to 4 days and happens quicker on high pH soils. Unless it rains, urea must be incorporated during this time to avoid ammonia loss. Losses might be quite low in the spring if the soil temperature is cold. The chemical reaction is as follows:

$$CO(NH_2)_2 + H_2O + \text{urease} \rightarrow 2NH_3 + CO_2 \text{ (urea)}$$

The problem is the NH_3, because it's a gas, but if incorporated the NH_3, acts the same as incorporated anhydrous ammonia. Also, half of 28 per cent liquid N is urea and the same thing happens with this half as with regular urea.

UREA LOSSES RELATED TO SOIL TEMPERATURE AND PH

The volatility of urea depends to a great extent on soil temperature and soil pH.

Table. Per cent of surface-added urea volatilized as ammonia at different temperatures and days on the surface.

	Temperature (F)			
Days	45 degrees	60 degrees	75 degrees	90 degrees
	(% of added N volatilized)			
0	0	0	0	0
2	0	0	1	2
4	2	2	4	5
6	5	6	7	10
8	5	7	12	19
10	6	10	14	20

Table. Per cent of surface-added urea volatilized as ammonia at various soil pH levels and days on the surface.

	Soil pH					
Days	5.0	5.5	6.0	6.5	7.0	7.5
	(% of added N volatilized)					
0	0	0	0	0	0	0
2	0	0	0	0	1	5
4	1	2	5	10	18	20
6	4	5	7	11	23	30
8	8	9	12	18	30	33
10	8	10	13	22	40	44

Fall application comparisons

Urea can be readily nitrified – that is, converted to nitrate (NO_3) – even when applied late in the fall, and can be quite susceptible to denitrification or leaching the following spring. Anhydrous ammonia (AA) applied in the fall does not nitrify as quickly, due to the stunting of microorganisms in the AA application band.

A two-year study conducted at Waseca compared late-October applications of both AA and urea for continuous corn. These data show a 6 bu/A advantage for AA over urea when applied in the fall without a nitrification inhibitor. But when N-Serve was added, a 16 bu/A advantage was shown with AA. This indicates that the inhibitor has a better degree of contact with the AA mix than is possible with urea.

Table. Corn yield as influenced by N source, time of application, and nitrification inhibitor at Waseca.Malzer and Randall.

N Source*	Fall	Spring
	Yield (bu/A)	
AA (82% N)	162	168
AA + N-serve	170	172
Urea (45% N)	156	164
Urea + N-serve	154	162

Note:

*150 lb N/A

Studies with continuous use of urea have been conducted at Lamberton since 1960. Corn yields over a 24-year period averaged 5 to 6 bushels per acre higher with spring application of urea compared to the fall plowed-down application.

Table. Corn yield as influenced by fall and spring applications of urea at Lamberton.

Time/method of urea application*	24-year avg. yield
	bu/A
Fall, plowed down	97
Spring, top-dressed	102
Spring, side-dressed	103

Note:

*80 lb N/A

Urea applied in the fall has generally not been as effective as AA. This is especially true in south-central Minnesota and Iowa. When fall soil-moisture conditions are dry, there is little difference between AA and urea. But when soil-moisture content is high, fall applications of urea haven't performed as well as AA.

Applications of urea-ammonium nitrate (UAN) in the fall are not recommended due to rapid nitrification and a high potential for loss.

SOIL APPLICATION AND PLACEMENT OF UREA

If properly applied, urea and fertilizers containing urea are excellent sources of nitrogen for crop production. After application to the soil, urea undergoes chemical changes and ammonium (NH4 +) ions form. Soil moisture determines how rapidly this conversion takes place. When a urea particle dissolves, the area around it becomes a zone of high pH and ammonia concentration. This zone can be quite toxic for a few hours. Seed and seedling roots within this zone can be killed by the free ammonia that has formed. Fortunately, this toxic zone becomes neutralized in most soils as the ammonia converts to ammonium. Usually it's just a few days before plants can effectively use the nitrogen. Although urea imparts an alkaline reaction when first applied to the soil, the net effect is to produce an acid reaction. Urea or materials containing urea should, in general, be broadcast and immediately incorporated into the soil. Urea-based fertilizer applied in a band should be separated from the seed by at least two inches of soil.Under no circumstances should urea or urea-based fertilizer be seed-placed with corn.

Table. Seed-placed ammonium nitrate (AN) and urea comparisons on seedling damage to spring wheat under limited moisture conditions. North Dakota, 1975.

Treatments		Seedlings per 40 ft. of row		
		Location		
N (lb/A)	N Source	Absaraka	Williston	Casselton
0	-	600	270	760
20	AN	570	220	600
30	AN	590	240	690
40	AN	590	260	660
20	Urea	400	200	550
30	Urea	280	110	430
40	Urea	220	70	220

Table. Effect of urea and ammonium nitrate placed with seed on corn grain yield.

	Yield, bu/A	
lb. N/A*	Urea	Ammonium nitrate
0	137	137
5	60	142
10	36	143
20	33	92

Note:

* Sufficient N broadcast prior to planting. Source: Liegel & Walsh Plainfield Sand, Hancock, Wisconsin

Table. Effect of urea and ammonium nitrate side-placed on corn grain yield.

	Yield, bu/A	
lb. N/A*	**Urea**	**Ammonium nitrate**
0	142	142
25	145	145
50	146	146
100	150	141

Note:

* Sufficient N broadcast prior to planting. Source: Liegel and Walsh Plainfield Sand, Hancock, Wisconsin

Table. Effect of source and placement of urea and ammonium nitrate (AN) on corn yields.

lb. N/A	Treatment	Source	Av. yield bu/A
0			62
40	Plow-down – fall	AN	79
40	Plow-down – fall	Urea	86
40	Surface – fall	AN	82
40	Surface – fall	Urea	85
80	Plow-down – fall	AN	98
80	Plow-down – fall	Urea	97
160	Plow-down – fall	AN	104
160	Plow-down – fall	Urea	105
40	Topdress – spring	AN	89
40	Topdress – spring	Urea	88
80	Topdress – spring	AN	100
80	Topdress – spring	Urea	102

With small grains, 10 lb. of nitrogen as urea can generally be applied with the grain drill at seeding time even under dry conditions. Under good moisture conditions, 20 lb. of nitrogen as urea can be applied with the grain drill. Research results at North Dakota State University indicate that under dry conditions, urea at the rate of more than 20 lb. nitrogen per acre, applied with a grain drill in a 6-inch spacing, can reduce wheat stands more than 50 per cent Research at the University of Wisconsin indicates that seed-placed urea with corn, even at low rates of nitrogen, is very toxic to the seed and greatly reduces yields. When urea was side-placed as a 2" x 2" starter, however, little if any damage was noted.

In Minnesota, good crop production usually requires an application of more than 20 lb. of nitrogen per acre. Farmers can avoid damage from urea by broadcasting most of the urea nitrogen fertilizer ahead of seeding. The urea broadcast prior to seeding is equal to or more effective than similar ammonium nitrate treatments.

SPREADING OF UREA

Urea can be bulk-spread, either alone or blended with most other fertilizers. It is recommended that the spreading width not exceed 50 feet when combined with other fertilizer materials.

Urea often has a lower density than other fertilizers with which it is blended. This lack of "weight" produces a shorter "distance-of-throw" when the fertilizer is applied with spinner-type equipment. In extreme cases this will result in uneven crop growth and "wavy" or "streaky" fields.

BLENDING UREA WITH OTHER FERTILIZERS

Urea and fertilizers containing urea can be blended quite readily with monoammonium phosphate (11-52-0) or diammonium phosphate (18-46-0).

Urea should not be blended with superphosphates unless applied shortly after mixing. Urea will react with superphosphates, releasing water molecules and resulting in a damp material which is difficult to store and apply.

Fluid urea

Uniformity of particle size is important with dry solid urea, whether applied directly or in blended formulations. Some imported urea appears to be below U.S. quality standards on granule uniformity. Dissolving urea and marketing the liquid solution is an attempt to overcome this lack of uniformity and still take advantage of the favorable urea price.

The liquid mix of urea and ammonium nitrate (UAN 28 per cent N) has been on the market for a long time. The characteristics of this solution, however, are not the same as when urea alone is dissolved in water. A solution of 50 per cent urea by weight results in 23-0-0 and has a salting-out temperature of 60 degrees F. In order to store and handle liquid urea during cooler temperatures, the nitrogen concentration must be lowered to reduce salting problems. There are several possible formulations that can be used for this, such as adding small amounts of ammonium nitrate, ammonium sulfate, or anhydrous ammonia.

Research, particularly on liquid urea, is very limited. Generally, where dry urea functions successfully, the fluid urea should perform equally well and may have the advantage of better uniformity over some dry urea sources.

Biuret in Urea

Biuret in urea can cause agronomic problems if placed near the seed. or even if added preplant in bands where seeds will later be planted.

Most U.S. manufacturers of urea keep biuret content low by keeping high temperatures to a minimum. Biuret content is typically around 0.3 per cent, although urea of foreign origin appears to be higher. High heat is normal during the manufacture of urea. If heat exceeds 200 degrees F there is a slight conversion of urea to biuret, but this takes place only during the manufacturing process. No such conversion happens in storage or in the soil.

Biuret converts to ammonia, but conversion is much slower than for urea. Since biuret remains in the soil for several weeks, the potential for seed damage continues beyond the brief period of conversion of urea to ammonia. The major damage of biuret is to germinating seeds. There is little damage through plant absorption, although some citrus crops have been affected.

Application of urea to growing crops

Urea can be applied to sod crops, winter wheat. or other small grains. This application, however, should be made during cool seasons. During warm periods (60 degrees F or above), urea in contact with vegetative material will tend to give off ammonia.

If urea must be applied on grass pastures in the summer, apply when there is a high probability of rainfall.

Foliar application of urea

Urea can also be applied as a foliar spray on some crops, such as potatoes, wheat, vegetables, and soybeans. Urea is highly water soluble. At normal atmospheric temperatures, approximately 1 lb. Of urea can be dissolved in 1 lb. of water.

Research data indicate that urea should contain no more than 0.25 per cent biuret for use in foliar sprays. For many crops the quantity of nitrogen applied at one time should not exceed 20 lb. of nitrogen per acre.

Urea storage

Urea is neither combustible nor explosive. It can be stored safely with no loss of quality under normal circumstances. Small or fast-moving augers should not be used to move granular urea. Urea particles are generally soft and abrasion can break the granules. Belt conveyers should be used whenever possible.

Urea should not be stored with ammonium nitrate. These materials, when in contact, rapidly absorb water when the relative humidity is above 18 per cent. The relative humidity at which urea and ammonium nitrate absorb moisture from the air.

Table. Critical relative humidities (CRH) of urea, ammonium nitrate, and a mixture of the two.

Material	CRH%
Urea	75.2
Ammonium nitrate (A.N.)	59.4
Urea + ammonium nitrate	18.1

Slow release of urea

Urea fertilizer can be coated with certain materials, such as sulfur, to reduce the rate at which the nitrogen becomes available to plants. Under certain conditions these slow-release materials result in more efficient use by growing plants. Urea in a slow-release form is popular for use on golf courses, parks, and other special lawn situations.

METHODS OF FERTILIZER APPLICATION

In order to get maximum benefit from manures and fertilizers, they should not only be applied in proper time and in right manner but any other aspects should also be given careful consideration. Different soils react differently with fertilizer application. Similarly, the N, P, K requirements of different crops are different and even for a single a crop the nutrient requirements are not the same at different stages of growth. The aspects that require consideration in fertilizer application are listed below:

1. Availability of nutrients in manures and fertilizers.
2. Nutrient requirements of crops at different stages of crop growth.
3. Time of application.
4. Methods of application, placement of fertilizers.
5. Foliar application.
6. Crop response to fertilizers application and interaction of N, P, and K.
7. Residual effect of manures and fertilizers.
8. Crop response to different nutrient carrier.
9. Unit cost of nutrients and economics of manuring.

Fertilizers are applied by different methods mainly for 3 purposes:

1. To make the nutrients easily available to crops,
2. To reduce fertilizer losses and
3. For ease of application.

The time and method of fertilizer application vary in relation to

1. The nature of fertilizer.
2. Soil type and
3. The differences in nutrient requirement and nature of field crops.

Application of fertilizers in solid form: It includes the methods like:

I *Broadcasting:* Even and uniform spreading of manure or fertilizers by hand over the entire surface of field while cultivation or after the seed is sown in standing crop, termed as broad casting. Depending upon the time of fertilizer application, there are two types of broadcasting:

A Broadcasting at planting and

B Top dressing.

A Broadcasting at planting: Broadcasting of manure and fertilizers is done at planting or sowing of the crops with the following objectives:

1. To distribute the fertilizer evenly and to incorporate it with part of, or throughout the plough layer and
2. To apply larger quantities that can be safely applied at the time of planting/sowing with a seed-cum-fertilizer driller. It is adopted with the following condition:
   i. When N-ous fertilizers like amm. Sulphate, Amm. Sulphate Nitrate, Concentrated organic manures, are to be applied to the soil deficient in N or where N is exhausted by previous crops like fodder, *Jowar*, F. maize.
   ii. When citrate soluble P-tic fertilizers like basic slag and dia-calcium phosphate, are to be applied to moderately acid to strongly acid soils.
   iii. When K-ssic fertilizers like Muriate of potash and potassium sulphate are to be applied in potash deficient soil.

B Top dressing: Spreading or broadcasting of fertilizers in the standing crop (after emergence of crop) is known as top-dressing. Generally, NO3 – N fertilizers are top dressed to the closely spaced crops like wheat, paddy. *E.g.*: Sodium Nitrate, Amm. Nitrate and urea, so as to supply N in readily available from the growing plants. The term side dressing refers to the fertilizer placed beside the rows of a crop (widely spaced) like maize or cotton. Care must be taken in top dressing that the fertilizer is not applied when the leaves are wet or it may burn or scorch the leaves. The top dressing of P and K is ordinarily done only on pasture lands which occupy the land for several years. In some countries, aero planes are used for fertilizer application in hill terrains where it is difficult to transport fertilizers and where large amount are to be applied because of severe deficiency and under following situations:

1. Where very small quantities of fertilizers are needed over large areas. *E.g.*: Micro nutrients.
2. When high analysis materials are applied.
3. When fertilizer application may be combined with insect control or some other air operation and
4. As a labour and time saving device.

II Placement: In this, the fertilizers are placed in the soil irrespective of the position of seed, seedling or growing plant before or after sowing of the crops. It includes:

1. Plough sole placement: The fertilizer is placed in a continuous

band on the bottom of the furrow during the process of ploughing. Each band is covered as the next furrow is turned. By this method, fertilizer is placed in moist soil where it can become more available to growing plants during dry seasons. It results in less fixation of P and K than that which occurs normally when fertilizers are broadcast over the entire soil surface.

2. Deep placement or sub-surface placement: In this method, fertilizers like Amm. Sulphate and Urea, is placed in the reduction zone as in paddy fields, where it remains in ammonia form and is available to the crop during the active vegetative period. It ensures better distribution in the root zone, and prevents any loss by surface run-off. It is followed in different ways, depending upon local cultivation practices such as:
   i. Irrigated tracts: The fertilizer is applied under the plough furrow in the dry soil before flooding the land and making it ready for transplanting.
   ii. Less water condition: Fertilizer is broadcasted before puddling which places it deep into the reduction zone.
   iii. Sub – soil placement: This refers to the placement of fertilizers in the sub-soil with the help of heavy power machinery. It is followed in humid and sub-humid regions where many sub-soils are strongly acid, due to which the level of available plant nutrients is extremely low. P-tic and K-ssic fertilizers are applied by this method in these regions for better root development.

III Localized placement: It refers to the application of fertilizers into the soil close to the seed or plant. It is usually employed when relatively small quantities of fertilizers are to be applied. It includes methods like:

Advantages:

a. The roots of the young plant are assured of an adequate supply of nutrients,
b. Promotes a rapid early growth,
c. Make early Intercultivation possible for better weed control,
d. Reduces fixation of P & K.

1. Contact placement or combined drilling or drill placement: It refers to the drilling of seed and fertilizer together while sowing. It places the seed and small quantities of fertilizers in the same row. This is found useful in cereal crops, cotton and grasses but not for pulses and legumes. This may affect the germination of the seed, particularly in legumes due to excessive concentration of soluble salts.

2. Band placement: In this, fertilizer is placed in bands which may be continuous or discontinuous to the side of seedling, some distances away from it and either at level with the seed, above the seed level or below the seed level. There are two types of band placement: It includes hill and row placement.
   a. Hill placement: When the plants are spaced 3 ft. or more on both sides, fertilizers are placed close to the plant in bands son one or both sides of the plants. The length and depth of the band and its distance from plant varies with the crop and the amount of fertilizer as in cotton.
   b. Row placement: When the seeds or plants are sown close together in a row, the fertilizer is put in continuous band on one or both sides of the one or both sides of the row by hand or a seed drill. It is practiced for sugarcane, potato, maize, tobacco, cereals and vegetable crops. Higher rates of fertilizers are possible with row placement than hill placement. For applying small amount of fertilizers, hill placement is usually most effective.
3. Pellet application: In this method, fertilizer (N-ous fertilizers) is applied in the form of pellets 2.5 – 5.0 cm. deep between the rows of paddy crop. Fertilizer is mixed with soil in the ratio of 1:10 and make into dough. Small pellets of a convenient size are then made and deposited in the soft mud of paddy fields. It increases the efficiency of N-ous fertilizers.
4. Side dressing: Fertilizers are spread in between the rows or around the plants. It includes i) application of N-ous fertilizers in between the rows by hand to broad row crops like maize, S.cane tobacco, cereals which is done to supply additional doses of N to the growing crop. ii) Application of mixed or straight fertilizer around the base of the fruit trees and done once, twice or thrice in a year depending upon age.

SULPHUR FERTILIZATION IN CROP PRODUCTION

Generally, plants require about a tenth as much sulphur (S) as nitrogen (N), but sulphur deficiencies restrict plant growth as surely and severely as nitrogen deficiencies. Canola and alfalfa are examples of high sulphur-using crops.

Sulphur is not mobile in the plant, so a continuous supply of sulphur is needed from emergence to crop maturity. A deficiency of sulphur at any stage of growth can result in reduced yields.

For the first half century or more of cultivation, sulphur deficiency was not a concern on most of our soils because a large pool of organic sulphur was made available as organic matter mineralized. Over time, the pool of organic sulphur has declined significantly, mostly due to use by crops, particularly high sulphur-using crops such as canola and alfalfa.

Sulphur deficiencies occur throughout Saskatchewan, particularly:

- For high sulphur-using crops;
- On sandy soils;
- On gray soils in the northern grain belt; and
- On low organic matter soils.

Table. Sulphur uptake by crops in lb. sulphur/acre.

Crop	Yield/Acre	Grain	Straw	Total Uptake	Sulphur Uptake (average)
Spring wheat	40 bu.	4 - 5	4 - 5	8 - 10	0.2 lb./bu.
Barley	80 bu.	6 - 8	6	12 - 14	0.16 lb./bu.
Oat	100 bu.	4 - 5	8 - 9	12 - 14	0.13 lb./bu.
Rye	55 bu.	4 - 5	10 - 12	14 - 17	0.28 lb./bu.
Canola	35 bu.	10 - 12	7 - 9	17 - 21	0.54 lb./bu.
Flax	24 bu.	5 - 6	7 - 9	12 - 15	0.56 lb./bu.
Pea	50 bu.	6 - 7	5 - 7	11 - 14	0.25 lb./bu.
Lentil	30 bu.	4 - 5	4 - 5	8 - 10	0.30 lb./bu.
Potato	20 tons	11 - 13	5 - 7	16 - 20	0.9 lb./ton
Alfalfa	5 tons	--------	--------	27 - 33	6.0 lb./ton
Grass	3 tons	--------	--------	11 - 14	4.2 lb./ton
Barley silage	4.5 tons	--------	--------	14 - 21	3.9 lb./ton

Sulphur deficiency symptoms

Sulphur is essential for many growth functions in plants including nitrogen metabolism, enzyme activity and protein and oil synthesis. Generally, sulphur-deficient plants have short and/or spindly stems and yellowing of the young (top) leaves. With nitrogen deficiency, yellowing affects the older, lower leaves first. Sulphur-deficient canola can also have purpling and upward cupping of young leaves, delayed and prolonged flowering, pale-coloured flowers and fewer, smaller pods.

Sulphur-deficient alfalfa, pea and other legumes may have reduced nitrogen fixation.

For cereals and forage grasses, yellowing of the newly-emerging leaves is an indicator of sulphur deficiency.

Baking quality of bread wheat

Adequate levels of sulphur are required for bread wheat production to optimize baking quality. Wheat fields deficient in sulphur require the addition of about 10 pounds of sulphate-sulphur per acre.

Yield

Substantial yield increases can be obtained by applying sulphur fertilizer to crops having sulphur demand that cannot be satisfied by soil sulphur supply. Sulphur fertilization may also produce earlier and more uniform maturity and higher oil and protein levels.

How much sulphur fertilizer to apply

Both nitrogen and sulphur are important in plant protein synthesis. The correct balance of nutrients is particularly important for high sulphur-using crops like canola and alfalfa.

Canola takes up nitrogen and sulphur in a ratio of about five to one (for example, 20 lb.S/acre for 100 lb. N/acre) to optimize protein, oil synthesis and yield. Applying high rates of nitrogen without sulphur can lead to a lower yield than if no nitrogen was applied.

Table. The effect of sulphate-sulphur on canola and barley seed yield and alfalfa dry matter yield on sulphur-deficient soils.

Fertilizer	Canola NW Saskatchewan (one site)	Canola NE Saskatchewan (average of four sites)*	Barley Northern Alberta (one site)	Alfalfa NE Saskatchewan (average of three years at one site)**
No fertilizer	13 bu./ac.	9 bu./ac.	33 bu./ac.	0.90 tonnes/ac.
Nitrogen (N) Phosphorus (P) Potassium (K) [no Sulphur (S) for canola and barley]	12 bu./ac.	1 bu./ac.	26 bu./ac.	1.22 tonnes/ac. (with P) 0.85 tonnes/ac. (with K)
Nitrogen, Phosphorus Potassium Sulphur	32 bu./ac. (23 lb./ sulphate-S/ac.)	21 bu./ac. (13 lb. sulphate-S/ac.) 25 bu./ac. (27 lb. sulphate-S/ac.	76 bu./ac. (20 lb. sulphate-S/ac.)	2.78 tonnes/ac. (with S) 3.19 tonnes/ac. (with P, K, S)

Sulphur fertilizers

There are three main types of sulphur fertilizer.

Sulphate-sulphur fertilizers contain sulphur in combination with other nutrients, such as nitrogen or potassium. Sulphate is readily available to growing crops and sulphate-sulphur fertilizers dissolve quickly. The most common sulphate-sulphur fertilizer sold in Saskatchewan is granular ammonium sulphate (20-0-0-24, 21-0-0-24, 19-2-0-22).

Ammonium sulphate can be blended with other granular fertilizers, but care should be taken to ensure the physical nature of the ammonium sulphate will allow the blend to remain uniform. Potassium sulphate (0-0-50-18 and other formulations) is also available and is well suited to legume crops such as alfalfa, where both potassium and sulphur are needed but not nitrogen. There are other fertilizer products containing some sulphate-sulphur, either in a blend or in a manufactured product. Elemental sulphur is not immediately effective for soils very deficient in sulphur, but may be a useful part of a long-term sulphur fertilizer management plan for soils low in plant-available sulphur.

Elemental sulphur fertilizers (0-0-0-90 to 99) are granular, with 90 to 99 per cent sulphur in the elemental form. Elemental sulphur cannot be directly used by plants. It must first be converted to sulphate-sulphur (SO_4^{-2} -S) by soil microorganisms.

In general, the finer the particle size and more thoroughly it has been mixed in the soil, the faster it will convert to sulphate. Broadcast applications of elemental sulphur tend to convert to plant-available sulphatc more rapidly than banded applications. When broadcast on the surface, the freeze-thaw and wet-dry cycles help disperse the granules and further reduce the sulphur particle size to allow specific soil microorganisms to convert the elemental sulphur to plant-available sulphate-sulphur. The conversion of elemental sulphur to plant-available sulphate may be very slow when placed as a band in cold, dry soils.

It is difficult to accurately predict the rate of conversion of elemental sulphur to plant-available sulphate as it is sensitive to a variety of soil and environmental conditions. Depending on the product, elemental sulphur should be broadcast applied about two years or more before the crop will need sulphate-sulphur to give time for the conversion to sulphate to occur. Increase in plant-available sulphate should be verified through a soil test. The slow conversion of elemental sulphur to plant-available sulphate can be an advantage. High rates of elemental sulphur can be applied once every few years to soils that require sulphur for all crops, or to long standing, high sulphur-using crops such as alfalfa and other forage crops.

In spite of high rates of elemental sulphur addition, it is recommended to apply starter application of sulphate-sulphur fertilizer when growing canola.

Elemental sulphur can be blended with some granular fertilizers. Contact your fertilizer dealer for details.

Fertilizers containing sulphur as thiosulphate, such as liquid ammonium thiosulphate (12-0-0-26), must also be oxidized by microbes in the soil to the sulphate form. However, the oxidation is rapid and recommended rates of ammonium thiosulphate can be applied the year sulphate-sulphur is required. Ammonium thiosulphate can be applied before, during or after seeding. However, when applied as foliar or dribble band to a crop, direct contact with plant leaves may cause leaf scorching. Follow manufacturers' instructions for blending.

Animal manure can provide sulphur along with other plant nutrients, but the sulphur content and balance with other nutrients is variable and should be determined through manure analysis along with soil testing to determine relative requirements. For example, some liquid swine manures have low contents of available sulphur relative to nitrogen, and crop responses to supplemental sulphur fertilizer have been observed on manured soils in field trials where sulphur deficiencies exist.

Irrigation water in Saskatchewan, may supply enough sulphur to meet the needs of irrigated crops. For example, water from the South Saskatchewan River contains three to five pounds of sulphate-sulphur per acre-inch of irrigation water. Soil and water testing is encouraged.

Timing and method of application

Sulphur fertilizer can be applied in a number of ways.

Table. Timing of various sulphur application methods for grain crops according to sulphur fertilizer form.

* *Conversion to sulphate depends on soil and climatic conditions*

Application method	Granular ammonium sulphate (20 or 21-0-0-24) (19-2-0-22)	Granular elemental sulphur (0-0-0-90 to 99)	Liquid ammonium thiosulphate (12-0-0-26)
Broadcast or dribble banded (liquid)	Before and after seeding	About two years or more before the plants need it*	Before crop emergence
Broadcast and incorporated	Before seeding	About two years or more before the plants need it*	Before seeding
Banded	Before seeding		Before seeding
Seed placed	Limited amount during seeding		Limited amount during seeding
Banded near the seed	During seeding		During seeding

A limited amount of sulphate fertilizer can safely be applied with the seed. The amount depends on seed type (*i.e.*, canola, cereal, etc.), amount and type of other fertilizers, such as mono-amonium phosphate, width of and distance between bands, soil type and soil conditions. Safe amounts of ammonium sulphate or ammonium sulphate plus nitrogen fertilizer should follow the guidelines for maximum amounts of seed-placed nitrogen according to the amount of nitrogen added as seed-placed ammonium sulphate plus the nitrogen added in any other seed placed nitrogen fertilizer.

Like nitrate-nitrogen, sulphate-sulphur is very mobile so top-dressed sulphate-sulphur fertilizer will be moved into the soil by rain. Unlike nitrogen, top dressed sulphur fertilizer is not subject to volatilization losses. On soils low in sulphur, application of sulphate-sulphur fertilizer can be effective up to the first flower stage of canola (or mustard), though the earlier the sulphur is top dressed the more effective it will be.

Due to its mobility, sulphate can also be leached below the rooting zone by high amounts of rain on sandy soils.

Table. Relative effectiveness of sulphate-sulphur fertilizer applied at at different growth stages on seed yield of canola (average of six sites).

Seed yield with applied sulphate-sulphur (bu./ac.)		
Fertilizer treatment	**13 lb./ac**	**27 lb./ac.**
Nitrogen* alone	3	
Nitrogen + pre-seed incorporated sulphur	19	22
Nitrogen + side-banded sulphur at seeding	19	22
Nitrogen + seedrow-placed sulphur	19	21
Nitrogen + surface-broadcast sulphur at bolting	15	17
Nitrogen + foliar-sprayed sulphur at bolting	16	18
Nitrogen + surface-broadcast sulphur at flowering	12	14
Nitrogen + foliar-sprayed sulphur at flowering	14	15

Conversions from imperial to metric:
lb./ac. x 1.12 = kg/ha
tons/ac. x 2.24 = tonnes/ha
lb. x 0.45 = kg
ton x 0.91 = tonne

Using sulphur fertilizer to lower soil pH in Saskatchewan soils

Applying high rates of sulphur fertilizers to lower pH of the soil is not highly effective. This practice is usually done to try to make iron, zinc or other nutrients more available to some horticulture crops. Additions of high amounts of sulphur fertilizers can also increase the amount of soluble salts (sulphates) in the soil causing the salinity levels to increase. The increase in soil salinity can reduce yields of sensitive crops.

UNDERSTANDING FERTILIZER AND ITS ESSENTIAL ROLE IN HIGH-YIELDING CROPS

High crop yields often come under scrutiny because of the fertilizer levels needed to produce such yields and because of the perception and reality of the potential environmental impacts of those inputs.

Yet, maintaining food production for the growing world population requires using new technology and intensifying production and management to grow more food on current cropland. Fertilizer is essential (for accomplishing this.

Misuse(of agricultural fertilizers have undoubtedly occurred, and its impact on the environment needs to(be minimized. But it's important to remember to compare the risk of fertilizer application with the benefits for food production.

The time has come to dispel misconceptions and myths about fertilizers and nutrients, and to convey a correct message to a world that's becoming increasingly urbanized and removed from what agricultural production is all about – providing healthy food.

FERTILIZERS' CONTRIBUTION TO THE FOOD SUPPLY

A survey of U.S. crop production estimated that average corn yields would decline by 40 per cent without nitrogen (N) fertilizer. Even greater declines would occur if other macronutrients, phosphorus (P) and potassium (K) were also limited. Numerous long-term studies have also demonstrated the contributions of fertilizer to sustaining crop yields. For example, long-term studies in Oklahoma show a 40 per cent wheat yield decline without regular N and P additions. A long-term study in Missouri found that 57 per cent of the grain yield was attributable to fertilizer and lime additions.

Similarly, long-term trials from Kansas show that 60 per cent of the corn yield was attributable to fertilizer N and P.

Few people appreciate that corn yields have continued to increase in the Corn Belt of the United States without a similar increase in N (the nutrient addition most associated with run-off-related water impairment). In fact, N-use efficiency has increased at least 35 per cent in the past 25 years (meaning less N fertilizer is now required to produce a bushel of grain). Remarkably, more corn is being harvested without increasing N fertilizer application rates. Some of this improvement has also come from modern genetics and improved agronomic management.

Is manure the answer?

Animal manure can provide a useful nutrient supply for growing crops. However, many people have the mistaken idea that manure has a special property for building soils. Manures contain no more nutrients than present in the animal feed that composes the manure. Similarly, manures don't produce any organic matter that wasn't initially in the animal feed. The digestion process doesn't produce any additional nutrients or organic matter.

This means that whatever organic matter or nutrients are present in field-applied manure are simply digested crops from somewhere else. The hay, grain or silage that's harvested to feed animals is simply taken from one field and then applied to another field after passing through an animal, with the inevitable loss of nutrients and carbon (C) to allow the animal to grow.

Animal manures rarely contain the essential plant nutrients in the proper ratio required for growing crops. In fact, certain types of manure application frequently result in imbalances and accumulation of nutrients in the soil that can pose an environmental risk. Composts and manures can be good nutrient sources, but their mineralization depends on complex interactions of both soil and environmental factors that are difficult to predict, which commonly results in a lower efficiency than fertilizer.

In an ongoing study in England (begun in 1840), applications of farmyard manure increased soil C and N to a greater extent than did fertilizer N. However, soil physical properties such as aggregate stability and(water infiltration improved the most in(the treatments receiving fertilizer (N. The N leaching that followed the(recommended manure application ((75 tons per acre per year) was almost twice that (from the fertilizer N treatment (250 N pounds per acre per year).

There will likely be more livestock and animal manure in the future, and these animals will consume more grass and (crops that must be fertilized, and produce manure that must be managed. But the animals themselves will not provide the needed new nutrients. Expanding urbanization means more organic waste and biosolids to manage. But resistance to applying these materials back on the land seems to be growing, and their land application is banned in many countrics.

Proper crop nutrition plays a vital role in maintaining the world's food supply. Use fertilizer appropriately to get the best results, and don't be afraid to speak out for farming practices that are such a benefit to humanity.

FALL APPLIED NITROGEN IN THE CORN BELT

Risks associated with fall nitrogen (N) applications fit into four categories: logistical, agronomic, environmental and economic.

Logistical risks

The primary reason growers apply N in the fall is the flexibility it allows in

N management. Fall is typically drier than spring and offers more days suitable for field work. Farmers considering shifting their N applications from fall to spring must carefully evaluate the additional workload spring requires. Many regular spring field operations must be performed within a short time period. If farmers plant corn too late, they can incur significant yield and economic losses. If the spring rush causes equipment to run over ground that's too wet, soils can become compacted, compounding yield and economic losses that can last several years.

Agronomic risks

Fall applications of N are, at best, equal to in-season applications of N in terms of effectiveness. They are seldom superior. They carry a higher risk of lower production arising from N loss between application and crop need. To minimize agronomic risk, university extension publications and industry professionals recommend applying it when daily soil temperatures are cooler and typically sustain below 50 oF. Nitrifying organisms that convert ammonium (NH+) N_4 to nitrate (NO_3-) are, for the most part, inactive until temperatures rise above 40 oF. At temperatures above this level, nitrification rates start out slowly but increase rapidly and can vary by soil. So by applying N in the fall when soils are cooler, better chances exist that enough N will still be present for corn production the next year.

Environmental risks

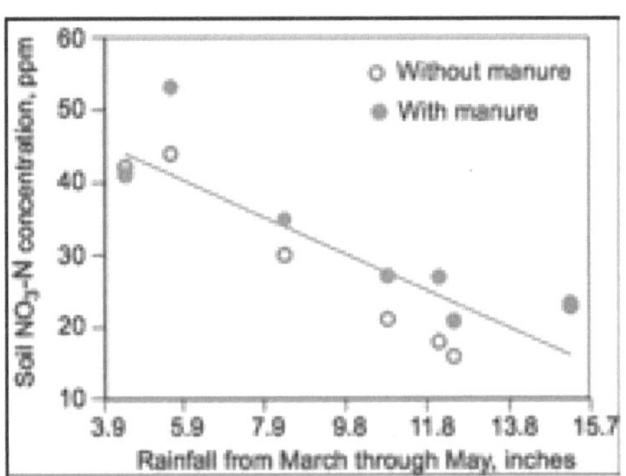

Variations within 0 to 12 inches of early-June soil NO3- levels across different years, and of March-through-May rainfall totals.

Compared with in-season applications, applying N in the fall increases the chances that leaching will cause some portion of the N to be lost prior to crop uptake.

In a multiple-year Iowa study, soil NO3-N concentrations in early June were well related to the amount of total rainfall received in March, April and May. As rainfall during this period increased, less NO3-N was present in the upper 12 inches of soil, indicating losses had occurred. This study illustrates that probabilities of N loss can be highly related to weather. Consequently, farmers can reduce the risk of N loss through leaching by applying N closer to the time of crop need.

If in-season N applications are rushed and soils become compacted, risks of N loss through denitrification (gaseous loss as N2O, NO or N2) may increase. The emission of nitrous oxide (N2O), a greenhouse gas, has been shown to be higher under compacted soil conditions. A study in Germany demonstrated that inter-rows compacted by wheel traffic emitted over six times as much N2O as non-compacted inter-rows and accounted for 68 per cent of the total N2O losses from the field.

Besides losing gaseous N, compacted soils can also reduce total N uptake by the crop. Lower uptake leads to yield losses and increases the quantity of N left in the soil after the cropping season. Depending on conditions, this unused N may be subject to losses.

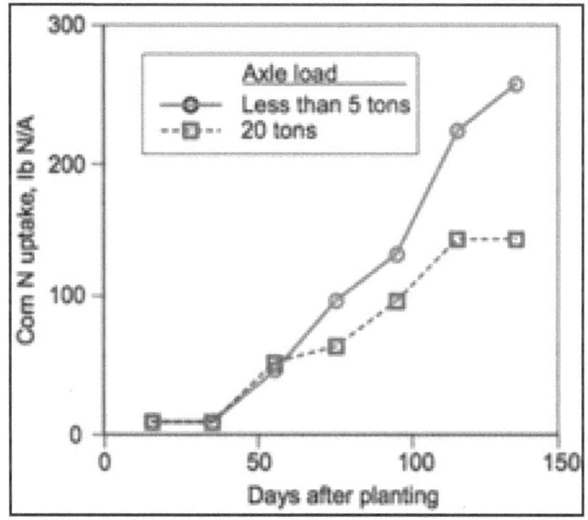

Effects of wheel-traffic-induced soil compaction on N uptake.

Appropriate N sources for fall

Nitrate sources. Sources of N containing NO_3- are not recommended for fall application for spring-planted corn. Soils don't hold in nitrate, so it moves with water through the soil profile.

Such N sources therefore run a high risk of NO_3- leaching and run-off losses(when crops are not present or actively growing for uptake. Common fertilizer products with such restrictions include urea ammonium nitrate (UAN) and ammonium nitrate.

Ammonium sources. Ammonium, however, is held by the soil and does not readily leach. Ammonium sources are therefore suitable for fall applications. Commonly applied forms of N in this category include ammonium sulfate, diammonium phosphate, and monoammonium phosphate.

Ammonia sources. Many consider anhydrous ammonia (NH_3) to be the best source to apply in the fall because it can delay nitrification. An early laboratory study showed that this effect depends on soil type and soil pH, among other factors. On a slightly basic Arredondo loamy fine sand, NH3 nitrified more slowly than ammonium sulfate. However, the reverse was true on a slightly acidic Lakeland fine sand. The temporary pH increases following NH3 applications may be favorable for nitrifying organisms at lower pH levels, but unfavorable at higher soil pH levels.

Table. Effeects of two N sources on NO_3-N and soil pH at two sampling dates on a slightly basic and a slightly acid soil.

Soil	N source @100 lb N/A	NO_3-N (percent of N applied)		Soil pH			
		14 days	28 days	Initial	After N addition	After 14 days	After 28 days
Arredondo loamy fine sand	Anhydrous ammonia	61	88	7.3	8.4	6.9	7.0
	Ammonium sulfate	87	92	7.3	7.1	6.8	7.0
Lakeland fine sand	Anhydrous ammonia	42	67	6.6	8.5	5.7	4.8
	Ammonium sulfate	21	39	6.6	6.2	5.0	4.6

Urea sources. Use of urea is acceptable for fall applications in most climates, if it's incorporated soon after application on soils with relatively low leaching and denitrification loss potentials. Once incorporated (by tillage or rainfall) and in the presence of adequate moisture, urea readily converts to NH + because of the action of soils. Movement of urea-N in soils is akin to NO_3- movement, so fall applications in sandy soils with a high leaching potential is generally not recommended.

Appropriate inhibitors for fall-applied N

Nitrification inhibitors. To limit denitrification losses, research recommends nitrification inhibitors in higher-rainfall areas of the Corn Belt with poorly drained soils and soils with higher moisture levels near the surface. Throughout the region, fall applications with a nitrification inhibitor have tended to be less effective in reducing the risk of losses at the same rate of N applied in the spring.

New evidence indicates that pH may also play an important role in determining effectiveness of nitrification inhibitors. December and February temperatures, effectiveness of a nitrification inhibitor was greater than in a year with colder December, February and March temperatures. In the warmer year, higher soil pH levels probably led to a more rapid recovery of nitrifying organisms from the inhibitor, resulting in a greater amount of nitrification at the time of measurement.

The influence of pH on the percentage of fall-applied N that had been nitrified by the following spring, with and without a nitrification inhibitor. Sub-figure "a" shows years with above-average spring precipitation, and sub-figure "b" shows years with below-average spring precipitation.

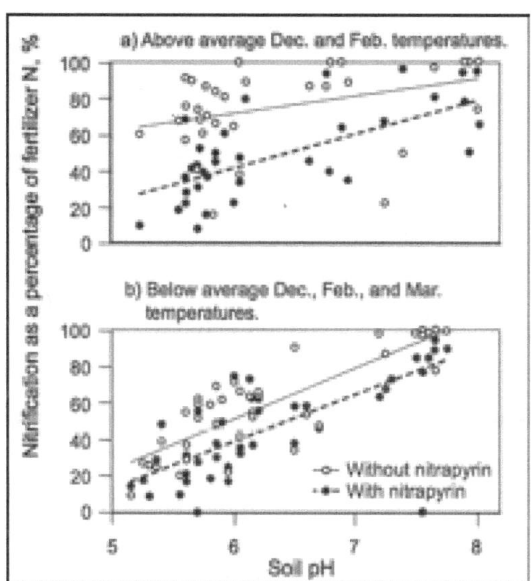

Urease inhibitors. For urea forms of N, a urease inhibitor may provide benefits when the incorporation of 0.3 to 0.5 inches of rain doesn't occur within 2 to 3 days after application.

5

Physical Aspects to Increase Crop Productivity

INCREASING OF CROP PRODUCTIVITY

Global demand and consumption of agricultural crops for food, feed, and fuel is increasing at a rapid pace. This demand for plant materials has been expanding for many years. However, recent increases in meat consumption in emerging economies together with accelerating use of grain for biofuel production in developed countries have placed new pressures on global grain supplies. To satisfy the growing, worldwide demand for grain, two broad options are available:

(1) The area under production can be increased or

(2) Productivity can be improved on existing farmland.

These two options are not mutually exclusive and both will be employed to produce the additional 200 million tonnes/year of corn (Zea mays) and wheat (Triticum aestivum) estimated to be needed by 2017. Both options will alter the environmental footprint of farming. Of the two options, increasing productivity on existing agricultural land is preferable as it avoids greenhouse gas emissions and the large-scale disruption of existing ecosystems associated with bringing new land into production.

In the United States, breeders, agronomists, and farmers have a documented history of increasing yield. U.S. average corn yields have increased from approximately 1.6 tonnes/ha in the first third of the 20th century to today's approximately 9.5 tonnes/ha.

This dramatic yield improvement is due to the development and widespread use of new farming technologies such as hybrid corn, synthetic fertilizers, and farm machinery. The introduction of biotechnology traits and development of new breeding methodology using DNA-based markers are further improving yields. Outside the United States, similar farming practices have been adopted in some agricultural nations, but in many major grain-producing countries, yields still lag well behind world averages. By continuing to develop new farming technologies and deploying of them on a global basis, demand for feed, fuel, and food can be met without the commitment of large land areas to new production.

SOIL PHYSICAL PROPERTIES AFFECT ROOT AND SHOOT GROWTH

Soil physical properties affect root and shoot growth directly and indirectly, the latter for example through poor drainage causing pores to fill with water and plants to suffer from anaerobiosis. Root growth has been described under various soil physical conditions, but relationships have only rarely been established between features such as crop yield, root growth and soil pore size distribution or conductivity, a more aggregate measure.

The difficulty in establishing simple relationships does not mean that soil has little influence upon root growth; rather it points to the complexity of the interactions and the internal homeostasis which plants maintain. Roots both elongate and proliferate and spread laterally as they grow and age. Concomitantly some roots die and others become suberised and function as conduits, but not as absorbers, of water and nutrients. Roots can elongate downwards as fast as 8 cm/d, as for example, soybean growing in a silt loam in a rhizotron.

Deep-rootedness and maximum rooting depth reflect soil properties (for example, roots will not grow through pores that they cannot deform to a larger diameter than the root).

However, relationships are not often reported. Maximum rooting depth varies with species and soil type. For example, wheat roots penetrated to 0.8 m in heavy-textured soils and to 1.2 m in a loamy sand but it is often found that a variety will have a consistent rooting depth across similar soil types in a particular year or in one soil across several years. Angus *et al.* found that rice and six dryland crops (mung bean, cowpea, soybean, groundnut, Maize and sorghum) extracted different amounts of stored soil water (ranging from 100 mm for rice to 250 mm for groundnut) and that extraction was, in part, related to rooting depth.

SPREAD OF ROOTS

The spread of roots with age can be related to the growth (increase in weight) of the whole plant, and to accumulated temperature or growing day-degrees (GDD); indeed, there is some evidence that temperature influences the direction of newly-appeared roots as well as the rate of appearance and extent of growth. Clearly, however, there are factors other than plant size, temperature and soil which influence root proliferation. Otherwise the plants sown at three different times of year in the same soil would align their root growth along a single growth-GDD relationship. These other factors, of which day length is probably particularly important, tend to mask underlying relationships between growth and soil structure.

The tillage can affect root length, though in this case its effects took three years to develop. Measurable differences in soil porosity developed under two tillage treatments: in the first year both root growth and water infiltration (K approximately 5 mm/h) were the same under minimum tillage and conventional tillage. By the third year, when differences were measured between roots, infiltration rates were 84 mm/h in minimum tillage and 0.2 mm/h under conventional tillage. Despite the differences in root growth there were no substantial differences in grain yield, reflecting the overall constraint of climate in the semi-arid environment.

One of the clearer associations between soil porosity and plant growth is described by Tisdall and Cockroft and Tisdall. Soil was ameliorated by deep ploughing, inserting gypsum at depth, and incorporation of straw and green manure in the topsoil. This treatment gave 10-fold increases in the number of earthworms and a 4-fold increase in soil pores. Infiltration rates also increased 10-fold relative to untreated soil. Root growth was not measured, but fruit yields from peach trees increased from 18 to 75 t/ha.

Increases in soil density or strength retard root penetration and thus limit the volume of soil exploited by the crop and the water available. It is difficult to quantify the relationships between these soil parameters and plant growth. In the cases of bulk density and strength, particularly, a gross measure of either for an undisturbed mass of soil can give only a remote indication of what a root encounters.

A determination of gross bulk density does not assess whether a root is growing within a pore (in which case it may deform surrounding soil before its radial environment reaches the density or strength of the gross soil) or if it is growing within the soil material, in which case it has already exerted a radial force equivalent to that measured for the gross soil.

This problem of scale-what is measured in a gross estimate cannot assess the micro-environment of the root, to which it (and through hormonal signals, the plant top) responds—does not invalidate some general hypotheses about plant behaviour in response to soil compaction and structural arrangement. Roots stop growing when they are unable to deform their micro-environment. This probably occurs at suctions about 60 kPa when they cannot generate adequate internal turgor. It is thought that root elongation declines curvilinearly with increasing either bulk density or shear strength. In the case of bulk density, little effect is noted until a 'critical value' and root elongation ceases within a further 10 per cent increase in bulk density. As explained earlier, these values vary with soil type. Root elongation declines asymptotically with increasing soil strength though the actual critical values would be expected to vary with soil type, water content and the method of measurement.

EFFECT OF BULK DENSITY

Studies of the effect of bulk density or strength on other plant processes, particularly germination and shoot elongation, also suffer from the same technical problem of scale. Shoots are able to explore macropores without being subject to the gross values of the soil they are in. The actual local values which inhibit shoot elongation appear to be quite small, for example, 0.76 kPa. These values from controlled situations contrast sharply with gross field values and with the large number of inconsistent correlations which arise from attempting to correlate crop performance with grossly-measured soil structure. Perhaps, most encouragingly, studies of genetic variation in the sensitivity of crops to soil strength suggest that there is appreciable range in plant sensitivity. The relativc ranking of genotypes is, however, the same when under near-critical stress as when growing with virtually no mechanical stress. Genotypes suited to stressful situations may be selected, therefore, by screening at a single soil strength.

The above comments relate to the direct effects of soil physical properties on crop growth. Two further points may be made. First, changing soil properties may not cause, or at least not immediately cause, measurable differences in plant performance.

This neither invalidates an association between the soil and the plant, nor the need to remain concerned about changes in soil properties. It is common but not often reported, that changes in a cropping system and soil characteristics first affect aspects other than the crop. It is only when a system has become significantly degraded, and there are associated environmental impacts, that reduced plant performance is noticed.

Nigeria, where no-till Maize in rotation with cowpea gave similar yields to conventionally cultivated crops. The no-till fields had only one-tenth the run-off and erosion of those conventionally cultivated. Likewise, alley-cropping of the annuals between six year-old hedgerows of tree legumes (*Leucaena leucocephala, Gliricidia sepium*) caused increased crop yields, but within the alley-cropping systems the yields of Maize were very similar despite substantial variations in run off and soil loss.

The second point about soil-crop relations is that the crop also affects the soil through ground-cover, depth of rooting and other properties. The crop attributes that most influence soil physical properties are speed of establishment and development of foliage cover.

Rapid establishment and growth minimizes topsoil structural decline and soil erosion by wind and water. Thereafter, deep-rooting directly affects soil structure, particularly if deep-rooted crops, such as safflower, are grown in rotation as a 'biological plough' to create macropores and these are minimally disturbed before the next crop is sown. A general list of desirable crop characteristics. Not all these are universally applicable (nor indeed, accepted by all scientists) because some attributes such as stolons and rhizomes, on the one hand, can provide advantageous ground cover and bind the soil, but they are undesirable if the species presents weeding or other problems.

The list includes both herbaceous and tree species. Ezenwa considers the attributes important for the choice of trees in Sahelian areas as the ability to:

- Enrich a microsite by depositing litter;
- Fix nitrogen and not be allelopathic;
- Avoid competition;
- Withstand stress, either water or wind;
- Provide alternative produce such as fodder, fuel or fruit.

BENEFITS OF COVER CROPS TO PRODUCTIVITY, TILTH, AND HEALTH OF SOILS

- Long-term soil health and productivity goals can be achieved with the use of cover crops.
- Cover crops are grown to improve soil nutrient levels, residue cover, and organic matter.
- Efficient use of water and fertilizer can be achieved when cover crops improve soil structure.

The long-term goals and benefits of soil health and productivity can be gained from the use of cover crops in rotations. Introducing cover crops into the rotation between corn and soybeans can improve soil 'tilth', health, and productivity. Tilth is a physical condition of soil and depends on aggregation-the joining of individual soil particles into clusters. Increased aeration and water-holding capacity is demonstrated in soils that have good tilth, drain well, resist crusting, and quickly take in water. When used long-term, cover crops provide a food source for soil organisms; in turn, soil organisms help with soil nutrient supply and plant uptake.

REDUCTION OF EROSION, RUN-OFF, AND NUTRIENT LOSS

Physical characteristics of a soil help determine the ability of a soil to resist erosion. Aggregates can be bound by a protein called glomalin, a water-insoluble substance produced by mycorrhizal fungi. Legume cover crops in particular, increase mycorrhizae populations, and can promote their relationships with subsequent crops. Glomalin could explain improved water infiltration and water storage on soils grown with cover crops even when increased organic matter levels have not been detected.

Erosion. Wind erosion depends on soil particle size, roughness of soil surface, wind speed, distance from shelter, and plant growth. Soil surface residue cover from no-till practices has been estimated to reduce wind erosion by 60 per cent compared to conventional tillage in Michigan.Soils are especially prone to wind erosion in the Great Plains states, and growing vegetation helps stabilize the soil better than dead or incorporated residue. Depending on geography, growers may want to consider the water use efficiency of cover crops grown long-term.

Some soil loss can occur without detectable yield loss; however, nutrients and water often need to be replaced at an expense to the grower. The USDA NRCS reported an average soil loss of 2.1 tons per acre per year from wind erosion in 2007. Soil that is moved off-site is often topsoil that contains organic matter and nutrients. While the organic matter and nutrient levels of soils vary, each ton of lost topsoil could contain 20 pounds of nitrogen (N), 12 pounds of phosphate, and 9 pounds of potash.

Fig. Fall growth of radish and Austrian winter pea cover crop mix. Radish has the ability to grow roots three feet into the soil profile, and is considered an excellent N scavenger. Austrian winter peas are great N providers with 90 to 150 lb. of N/A contributed by their growth.

Run-off Intense thunderstorms that last a short time cause noticeable soil movement. The impact of rain drops and shear force of water flow leads to inter-rill and rill erosion, respectively. Rills are shallow incisions into the topsoil and are often the first sign of erosion on exposed slopes. Soils are especially vulnerable to run-off and erosion during spring months when soils are often saturated, snow is melting, and vegetation is minimal. Run-off leads to nutrient loss, soil loss, and off-site contamination to surface waters. Soil water recharge is also reduced when precipitation moves off-site. Cover crop growth and residue can slow or stop run-off.

The canopy of cover crops shield the soil and reduces the impact of rainfall energy. Shear force of water flow can be strong during periods between cash crops. Cover crops reduce the speed of water flow over the soil surface and keep surface residue in place.

Run-off issues can be complicated by soil compaction. Subsoil compaction can be broken up or 'bio-drilled' by deep-rooted cover crops such as forage radish or annual ryegrass. Soil structure is left intact when a cover crop, rather than tillage, is used to improve soil structure issues. Tillage quickly negates benefits of cover crops as soil structure is broken down with the use of machinery. Soils may be re-compacted by machinery, sometimes even worse than before deep tillage.

Nutrient Retention. Nutrient loss from run-off, erosion, and leaching can be slowed or stopped with cover crop root and canopy growth. Nitrate-N is the nutrient frequently lost with water as run-off or leached through the soil profile. Cover crops take up or sequester excess N and reduce leaching into groundwater. Deep profile nutrient scavenging can draw nutrients up through the soil profile and release them during cover crop degradation. Subsequent cash crops can take up the released nutrients.

Productivity

Nitrogen Fixation. A 2013 survey conducted by the Conservation Technology Information Center (CTIC) and the USDA found that 41 per cent of the growers were planting cover crops to scavenge for N. Legumes that fix and release N for a cash crop to recapture in the spring demonstrate the value to cover crops. Cover crop seed cost and ability to supply potential N should be compared to fertilizer costs. The Midwest Cover Crop Council (MCCC) web site can help select cover crops and estimate how much N is returned.

Residue Cover. 'Soil erosion reduction' was a cover crop characteristic sought by 56 per cent of respondents in the 2013 CTIC and USDA survey. Along with decreased erosion and run-off, early spring growth of cover crops can help use excess moisture and stabilize the seed bed. This stability can allow machinery to pass over cover-cropped fields sooner compared to tilled fields. Termination of spring cover crop growth should be timed early enough to preserve adequate water for the cash crop. Herbicides, mowing, or another method besides tillage should be used to terminate cover crops. Maintaining cover crop roots such as in no-till fields, helps provide year-round habitat and food sources for microorganisms. Microorganisms are key to nutrient availability and pest suppression.

Restoration. Cover crops can be used to restore lightly degraded soils over time. Restoration of soil productivity is important for sustainable agriculture as yields reach their full potential when soils have good tilth and health. Cover crops can help restore marginally productive acreage. Slight to complete crop loss can result from salinization - the accumulation of water-soluble salts in the root zone of crops. This type of soil degradation is caused by irrigation, or poor soil-water drainage where there are naturally-occurring soil and water salts. Salinization is complex; however, it can be corrected by removal of salts from the root zone. Salt-tolerant cover crops, such as barley or sugarbeet, can be used to increase transpiration and reduce evaporation that leads to salt accumulation in the root zone.

Prevented Plant Acreage. Some planting seasons end with millions of acres either left unplanted or enrolled as prevented plant acreage. Soil left fallow lacks root growth, and mycorrhizal fungi that promote phosphorus (P) uptake do not survive flooded soil conditions or without host roots. Corn crops planted the year after flooding may have plants with P deficiency symptoms, also known as 'fallow syndrome'. Mycorrhizal growth is important for supplying P to early corn growth and correction of 'fallow-syndrome'. Cover crops can favour growth of mycorrhizal fungi and increase P levels within the plant comparable to if P was applied with the corn seed at a rate of 6.3 lbs/ac.

RESTORING THE PRODUCTIVITY OF MARGINAL SOILS WITH ORGANIC AMENDMENTS

Regular additions of organic materials such as animal manures and crop residues are of utmost importance in maintaining the tilth, fertility and productivity of agricultural soils, protecting them from wind and water erosion, and preventing nutrient losses through run-off and leaching. These materials have predictable beneficial effects on soil physical properties such as increased water-holding capacity, soil aggregation, soil aeration and permeability, and decreased soil crusting and bulk density.

Failure to recycle organic wastes and residues, intensive row crop production, and lack of sod-based crop rotations can result in extensive soil degradation and a decline in productivity due to excessive soil erosion and loss of fertility. The continuing desertification of Sub-Saharan Africa and the dustbowl of the central U.S. Great Plains in the 1930's can be attributed largely to improper farming methods that neglected the importance of soil organic master in crop production.

When organic materials, such as compost, animal manures, crop residues and sewage sludges are used as the primary sources of plant nutrients, the management system has often been referred to as "organic farming". More recently, the terms alternative, regenerative. low-input, and sustainable have been used to describe farming systems that recycle available on-farm organic resources and sometimes off-farm materials such as municipal wastes, to maintain or improve soil productivity.

In addition, organic materials can be used effectively for land reclamation purposes. For example, the mining of topsoil and sand and graver deposits in urban areas has left extensive tracts of exposed, highly erodible subsoils which are not conducive to the support of plant growth because of their adverse chemical and physical properties. Such areas detract from the aesthetic value of an urban environment and are major contributors to environmental pollution through surface run-off, eutrophication (*i.e.*, nutrient enrichment) of lakes and streams and sedimentation from soil erosion. Thus, there is a need to develop sound practices for restoring the productivity and value of these lands in the most economic and expedient way.

A wide variety of different organic wastes and residues can be used as soil conditioners and sources of plant nutrients on agricultural soils, with little or no adverse effects on public health and the environment. The purpose of this chapter is to discuss how some organic amendments can provide an effective means of restoring the productivity of marginal, degraded, and infertile soils and how changes in cultural practices may affect crop quality.

RECLAMATION OF DEGRADED AND MARGINAL SOILS

Some of the areas mined for sand and gravel in the northeast corridor of the United States are owned by small farmers who, after mining operations are completed, want to revegetate the barren areas to minimize erosion and improve the aesthetic value of the land. Som~ want to put the land back into agricultural production so as to gain a more favorable tax assessment.

Research was conducted by the USDA Agricultural Research Service beginning in the late 1970's to determine the feasibility and practicability of re vegetating these spoiled lands. After mining, the residual spoil materials were generally acidic (pH 4.5), of sandy texture and lacking sufficient organic matter and nutrients to maintain plan growth. For two years, 0, 40, 80 and 16 mt/ha of feedlot manure and sewage sludge compost produced by the Beltsville Aerated Pile Method were applied to spoil area plots and incorporated by rototilling.

All treatments were replicated three times. After two cropping years with sweet corn and two repeated applications of each waste, both the soil organic master and gravimetric water content increased significantly. Additions of 160 metric tons per hectare of either compost or manure increased the organic master content from an initial lever of 0.8 per cent to 4.2 per cent and 8.5 per cent, respectively. Gravimetric water content was similarly affected, increasing from 7 per cent for the control soil, which received only the recommended rate of commercial inorganic fertilizer for sweet corn, to 15 per cent and 32 per cent for the 160 mt/ha rate of compost and manure, respectively. Initially the manure contained twice as much water (80 per cent by weight) as did the compost (40 per cent by weight), but the larger increase in the soil water content for the manure-treated plots versus the compost-treated plots was attributed to the fact that the manure had a higher water-holding capacity than did the compost.

These changes in soil physical properties are important, because the increased lever of organic master increased the soil's capacity to retain more plant nutrients, including water, resulting in increased crop growth and yield. During drought conditions, sweet corn grown on the manure-treated plots showed fewer signs of moisture stress (wilting and rolling of leaves) than did the compost-treated plots, again indicating the higher water-holding capacity of the manure over compost.

With sandy soils, high afternoon temperatures can dry out newly planted seeds and, therefore, reduce cotyledon emergence and seedling survival. Soil temperatures taken at a 4-inch depth on plots receiving either 160 mt/ha of compost or manure were consistently 5 to 10 degrees Fahrenheit longer than on the unamended control soils. The magnitude of temperature lowering was directly related to application rate of the organic material. Such a reduction in temperature may not be desirable in heavy soils in cool regions because of possible adverse effects on seed germination. There was considerably less erosion on those plots treated with manure or compost because of an improvement in soil physical properties, particularly soil structure. This was evidenced by visible gullying in the control plots.

Significant increases in corn stalk yields were noted for plots treated with compost and manure. However, sweet corn ear yields were not always improved by additions of organic materials. Sahs and Lesoing (1985) also noted similar results when corn was grown with synthetic fertilizer versus feedlot manure. For corn grown on manure-treated plots they found significantly higher yields in a drought year, but significantly longer than normal yields in years of average temperature and adequate rainfall. Similar results and observations were also cited by the USDA study team on organic farming.

Table. Crop yield response to sewage sludge compost applied to a sand and gravel spoil site.

Sludge Compost applied (mt/ha)	**Field corn stalks (kg/ha)**	**Bush bean pods (kg/ha)**
0	6092	878
40	–	1175
80	6819	1309
160	7952	–
Treatment (T/ha)	**Organic Matter (%)**	**Soil Moisture (%)**

Control		0.8	7
Compost	40	1.5	10
	80	2.5	12
	160	4.2	15
Manure	40	3.1	15
	80	2.8	19
	160	8.5	32

The most dramatic increase in crop yield occurred when green snap bush beans were grown on plots receiving 0, 40 and 80 mt/ha of sewage sludge compost. The yields of green beans receiving 40 and 80 mt/ha of compost were 34 per cent and 49 per cent higher, respectively, than the yields on the control plots. In these studies, as well as those discussed earlier for sweet corn, the productivity of the sand and graver spoils was restored. to a relatively high lever in two to three years.

SOIL MANAGEMENT AND ENVIRONMENTAL CONSIDERATIONS

An important concept that is often overlooked is that for most agricultural soils, degradative processes such as soil erosion, nutrient run-off losses, and organic master depletion are going on simultaneously with conservation practices such as residue management, crop rotations, and conservation tillage. The potential productivity of a particular soil then depends on the interaction of degradative processes and conservation practices. On our best agricultural soils, *e.g.*, gently-sloping, medium-textured, well-structured, and deep-profile, a high level of productivity can be maintained by a relatively few, but essential, conservation practices that can readily offset most degradative processes. However, on marginal soils, *e.g.*, steeply-sloping, coarse-textured, poorly-structured, shallow-depth, and low fertility, soil conservation practices must be maximized to offset further degradation. The vital component in this dynamic equilibrium is organic matter.

Organic matter can be supplied to marginal soils from a number of sources including (1) on-farm wastes such as animal manures and crop residues, as well as green manure crops or (2) off-farm wastes such as sewage sludge and composted refuse. Off-farm wastes can be important sources of organic matter to farmers who do not have sufficient animals or cropping area to produce the amount of manure and crop residues needed to maintain soil productivity.

The plant nutrient contents of most organic materials are generally much lower than those supplied by commercially available chemical fertilizers. For example, the macronutrient content of crop residues can range from 0.7 to 2.5 per cent for nitrogen, 0.07 to 0.2 per cent for phosphorus and 0.9 to 1.9 per cent for potassium; animal manures can range from 1.7 to 4 per cent for nitrogen, 0.5 to 2.3 per cent for phosphorus, and 1.5 to 2.9 per cent for potassium: while sewage sludges range from 3 to 7 per cent for nitrogen, 1 to 3 per cent for phosphorus and 0.2 to 3 per cent for potassium.

Organic materials can differ widely in their properties and characteristics. Some materials, such as uncomposted animal manures, green manures, and sewage sludges, are subject to rapid microbial decomposition (*i.e.*, mineralization) in soils and tend to release their plant nutrients rapidly. This is desirable for soils that are already at a relatively high lever of fertility and productivity. On the other hand, some other materials, such as cereal straws, wood bark, and composted animal manures and sewage sludges, would be more resistant to microbial attack and release their nutrients at a relatively slower rate. This higher lever of organic stability provides a distinct advantage in the initial reclamation of marginal soils because it imparts a beneficial and long-term residual improvement of soil physical properties. Unless the physical nature of these soils is improved first, the plant Use efficiency of nutrients, whether from organic amendments or chemical fertilizers, will be unacceptably low.

In many developing countries there is a scarcity of suitable organic materials for composting or direct recycling on agricultural lands because of competitive uses. Thus, treating the entire soil-root zone with an organic amendment is often not feasible because of limited amounts of material. Consequently, the farmer must seek colt-effective methods of utilizing these materials to enhance soil productivity and crop yields. It is noteworthy that the productivity of marginal lands can be improved substantially with relatively small amounts of materials. This can be accomplished through localized placement techniques such as side-dressing, banding, bed and furrow systems, vertical mulching and slot-mulching.

Composting or co-composting an organic waste results in a more stable product that is easier to store and use. Co-composting of wastes that vary widely in their carbon to nitrogen ratios or solids content may produce a higher quality product and allows recycling of some wastes that could not be utilized as the only source of organic master due to some inherent chemical or physical property. Possible wastes that could be co-composted are municipal refuse or garbage, pit latrine wastes, sewage sludges, animal manures and crop residues.

Fertilizer use efficiency can be very low in strictly monoculture systems or where organic recycling is not practiced. This inefficiency allows for the movement of nutrients through the soil profile and into the "round water. Due to heavy use of pesticides and nitrogen fertilizer, contamination of "round water is evident in our major agricultural areas.

The possibility of Federal and State regulatory agencies enacting legislation to control these chemical inputs is also a major concern to the agricultural community.

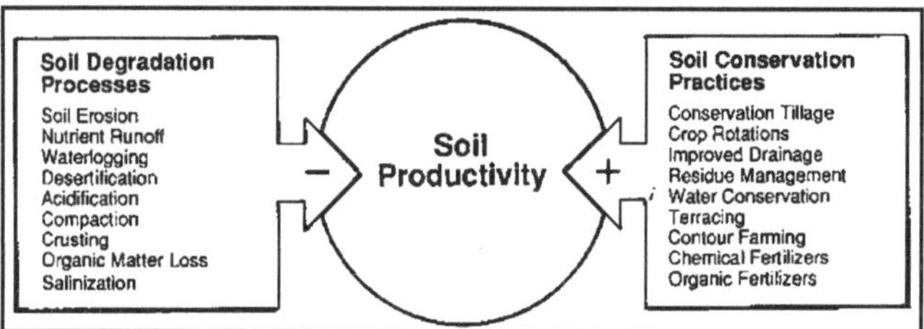

Fig. Relationship of soil productivity to soil degradative processes and soil conservation practices.

Many farmers have already pulled back from their long-held goal of achieving maximum yields because of the costly production inputs and low market prices which have steadily decreased their profitability. The recycling of organic wastes can longer production inputs by decreasing the amount of chemical fertilizer required for crop growth with minimal adverse environmental impact. As this change in cultural practice occurs, the effect of various longer fertilizer levers on crop quality could become very important.

CROP QUALITY AND NUTRITIONAL CONSIDERATIONS

Although there are many factors that can affect crop quality, the cultivar and post-harvest handling are considered to have the greatest effect on the nutrient composition of crops. One is reminded, however, that cultivars are often selected for their response to production inputs, especially chemical fertilizers, and with maximum crop yields as the primary consideration. As farmers attempt to reduce their dependency on chemical fertilizers and pesticides, they will undoubtedly adopt various cultural practices to fulfill the plant nutrient requirement and to control weeds and insects. Thus, cultural practices could have considerable impact on crop quality both now and in the future.

There has been much speculation on the benefits of consuming crops grown conventionally with chemical fertilizers compared with those grown on soils amended with organic materials as the primary source of plant nutrients. For example, researchers have shown conflicting results with respect to the ascorbic acid (vitamin C) content of crops.

Kansal et al. (1981) showed that as the lever of farmyard manure increased, the ascorbic acid content of spinach decreased. On the other hand, Schuphan (1974) reported higher ascorbic acid values for spinach grown on manure treated soil compared with chemical fertilizer. Similarly, Harwood (1984), using Pak Choi, and Ahrens et al. (1983), using spinach, also showed higher levers of ascorbic acid with increasing rates of composted manure versus chemical fertilizer.

The results of an NPK factorial experiment conducted to determine whether the results observed for ascorbic acid were due mainly to an organic amendment, or simply a fertilizer effect based on the nitrogen content of each amendment showed a significant decline in ascorbic acid content of kale as the rate of inorganic nitrogen increased. These results indicate that increased nitrogen rates do tend to depress ascorbic acid levels. While this may be attributed in part to a dilution effect from increased yields, other biochemical interactions appear to be involved.

Most of the conflicting results that have been reported on crop response to chemical and organic fertilizers are most likely due to the fact that the organic materials were mineralizing at different rates, thus, providing different levers of inorganic nitrogen to the crop. More-over, it is possible that results were influenced by the predominance of a particular form of inorganic nitrogen *i.e.*, nitrate (NO_3) or ammonium (NH_4), that was taken up by the plant. In a review of possible fertilizer effects on crop quality, Grunes and Allaway (1985) noted that the addition of boron to boron deficient soils increased the carotene (a vitamin A precursor) content in carrots and speculated that fertilizers that correct chlorosis in plants may increase the carotene content of the plant.

This concept is especially interesting in light of the fact that after basic needs for protein and calories are satisfied, the lack of a suitable source of vitamin A is one of the biggest health problems in developing countries. Perhaps with the addition of boron or through some other management factor we may also increase the lever of carotene or some other nutrient that is presently deficient in native diets.

Recently we have become increasingly concerned about "round water quality and the increased levers of nitrates and pesticides that have been reported in our major agricultural areas. Some states have already developed strategies for reducing the recommended rates of fertilizer nitrogen to minimize the risk of "round water pollution. It would indeed be ironic if it was shown that longer rates of fertilizer N were related to crops of a higher nutritional value.

BIOAVAILABILITY

The concept of producing better quality crops, however, is a complex issue because bioavailability of crop nutrients depends on many factors. Bioavailability refers to the amount of a particular nutrient that is absorbed from a food after consumption that is utilized by an animal or human It is not the total amount of a nutrient in the food that is consumed.

Measurements of nutrient bioavailability are difficult because of the many interactions that occur between minerals, vitamins and food components such as fibre. For example, iron bioavailability is determined by the iron status of the individual and the source of - iron being consumed.

It is enhanced by the presence of ascorbic acid (vitamin, C) in the meal. In addition, iron bioavailability may decrease when a meal high in dietary fibre is consumed by an individual who does not normally consume a large amount of dietary fibre. Therefore, each of these nutrients, iron, ascorbic acid, and fibre is important in the diet for optimum health and normal bodily function, but the presence of one might enhance or impair the function of the other.

In addition to these types of interactions, little is known about the bioavailability of nutrients in crops grown under different management conditions. At the 1st International Symposium on Horticulture and Human Health held in 1987 in Virginia, it was very apparent how little the horticulturalists and nutritionists knew about the practices and needs of one another.

It is clear that research on cultural practices and bioavailability of nutrients in foods and feeds consumed by humans and animals is needed, and should be conducted by multidisciplinary teams including soil scientists, agronomists, horticulturalists, crop breeders, physiologists, and nutritionists.

INCREASING CROP PRODUCTIVITY TO MEET GLOBAL NEEDS FOR FEED, FOOD, AND FUEL

Global demand and consumption of agricultural crops for food, feed, and fuel is increasing at a rapid pace. This demand for plant materials has been expanding for many years. However, recent increases in meat consumption in emerging economies together with accelerating use of grain for biofuel production in developed countries have placed new pressures on global grain supplies. To satisfy the growing, worldwide demand for grain, two broad options are available: (1) The area under production can be increased or (2) productivity can be improved on existing farmland. These two options are not mutually exclusive and both will be employed to produce the additional 200 million tonnes/year of corn (*Zea mays*) and wheat (*Triticum aestivum*) estimated to be needed by 2017.

Both options will alter the environmental footprint of farming. Of the two options, increasing productivity on existing agricultural land is preferable as it avoids greenhouse gas emissions and the large-scale disruption of existing ecosystems associated with bringing new land into production. In the United States, breeders, agronomists, and farmers have a documented history of increasing yield. U.S. average corn yields have increased from approximately 1.6 tonnes/ha in the first third of the 20th century to today's approximately 9.5 tonnes/ha. This dramatic yield improvement is due to the development and widespread use of new farming technologies such as hybrid corn, synthetic fertilizers, and farm machinery.

The introduction of biotechnology traits and development of new breeding methodology using DNA-based markers are further improving yields. Outside the United States, similar farming practices have been adopted in some agricultural nations, but in many major grain-producing countries, yields still lag well behind world averages. By continuing to develop new farming technologies and deploying of them on a global basis, demand for feed, fuel, and food can be met without the commitment of large land areas to new production.

Global demand for corn and wheat is growing at a rapid pace. As disposable incomes have risen in developing countries, meat consumption has increased. Among urban Chinese, meat consumption rose from 25 kg person"1 year"1 to 32 kg person"1 year"1 between 1996 and 2006. It is anticipated that meat consumption will continue to grow in developing countries because global consumption levels remain far below the approximately 100 kg person"1year"1 meat consumption rate of the United States and many western European countries.

Globally, meat consumption is expected to grow by 55 million tonnes to 310 million tonnes/year over the next decade. During this same period, biofuel production from corn and, to a lesser extent, wheat is expected to grow by 28 billion litres to 67 billion litres/year. Meeting the expected demand for meat will require feed grain usage to increase by about 50 million tonnes to about 640 million tonnes/year. Concomitantly, grain consumption for biofuel production is likely to increase by about 60 million tonnes to about 145 million tonnes/year. When food use for corn and wheat is added to the calculation, total demand for corn and wheat over the next decade is expected to increase by about 15 per cent or about 200 million tonnes/year to a total of approximately 1.5 billion tonnes/year.

The Food and Policy Research Institute (FAPRI) estimates that an additional 6 million ha of corn and 4 million ha of wheat plus a roughly 12 per cent increase in global corn and wheat yields will be used to produce this additional 200 million tonnes of grain. Both increases in planted area and increases in yield are likely to be needed to meet global demand for grain. However, improving yield on existing agricultural land will have a lower environmental impact than bringing new land into production. Cultivation of new acreage requires land clearing and subsequent tillage that results in significant greenhouse gas emissions also has negative impacts upon biodiversity and water quality.

Increasing the productivity of existing agricultural land will also have environmental consequences, but the negative consequences are generally less onerous and in some cases can be positive, depending upon how the land was previously used. Increased use of nitrogen fertilizers, a concern with both methods of increasing production, can increase nitrous oxide emissions, reduce water quality, and increase the size of hypoxic zones. However, incremental yield increases can be achieved on existing agricultural land through conservation tillage or transgenic insect control. Conservation tillage can decrease erosion, conserve soil moisture, and increase soil organic matter, and transgenic insect control can reduce broad spectrum insecticide use.

Global corn yields average 4.9 tonnes/ha and have been increasing steadily for many years. This is encouraging, but yields in major grain-producing countries are nearly double the global average, suggesting that there is room for significant improvements in global yields. Average corn yield in the United States is 9.4 tonnes/ha and Canadian farmers attain average yields of 8.2 tonnes/ha with this crop of tropical origin. In contrast, corn yields in the 10 largest lower-yielding corn-producing countries are just 2.8 tonnes/ha, well below the global average.

Much of the disparity in yields can be ascribed to agronomic practices, such as the use of open-pollinated corn varieties instead of hybrids, low input rates, or poor soil management. Brazil (29 per cent), India (56 per cent), and Romania (57 per cent) all plant significant amounts of open-pollinated varieties. Weather is also a significant factor in some countries, but the use of less robust production systems can magnify the effects of unfavorable weather in countries such as South Africa and Romania. Increasing the use of modern farming practices in these countries, together with the infrastructure, marketing, and risk management tools needed to support them, could lead to significant increases in crop production that limit the need to bring incremental land into production.

Higher crop prices, prompted in part by rising demand, have increased costs for urban consumers, especially those in poorer countries. However, higher crop prices will also provide farmers with the economic incentive to invest in farming methods and technologies that improve crop yields. Raising corn yields in the 10 largest, below average, corn-producing countries to just the world average will result in the production of an additional 100 million tonnes of corn or about 80 per cent of the projected growth in demand by 2017. Implicit in this scenario is the idea that rising yields will markedly diminish the global need for new crop acreage.

Rates of gain for yield have changed as new agricultural technologies have been developed and adopted. Average annual corn grain yields in the United States were relatively steady at approximately 1.5 tonnes/ha prior to the 1930s. Yields began to increase when hybrid corn was first introduced and the rate of gain accelerated further in the 1950s as single cross hybrids were introduced. Similarly, U.S. sorghum (*Sorghum bicolour*) yields increased sharply in the 1950s as hybrid sorghums were adopted. Since that time, yield has improved steadily because of fertilizer management, the development of more efficient farm machinery, and the breeding of hybrids with improved stress tolerance that in turn enabled higher plant populations and earlier planting dates.

Genetic gain experiments, in which hybrids that were widely grown at different points in time, often referred to as era hybrids, are compared side by side in the same trials, have been employed to estimate that about 50 per cent of the yield gained between the introduction of hybrid corn, and today is derived from breeding and the remainder from improved agronomic practices. Similar results have been reported from studies in France, Canada, and Brazil. As new farming technologies are adopted by farmers, the gap between test plot results, such as those reported by Duvick, and on-farm yield averages decreases. Between 1935 and 1990 this gap shrank from about 3.0 tonnes/ha to about 1.8 tonnes/ha. However, as rates of gain derived from breeding have increased in recent years, this gap appears to have widened again. This observation supports the hypothesis that the immediate future will witness a shift in average rates of gain as newer hybrids are adopted more widely in the United States and elsewhere.

Marker-assisted breeding and biotechnology traits are relatively new technologies for the improvement of productivity. Incorporation of these technologies into crop improvement programmes is likely to increase rates of gain beyond those seen in the last few decades. Results from a Monsanto Company study of a large number of commercial corn and soybean (*Glycine max*) populations indicated that use of markers can improve the rate of gain for yield and associated traits such as grain moisture and stalk lodging. Likewise, the suite of biotechnology traits currently used in commercial production in the United States increases average yields by protecting corn from the stress of competing pests and weeds.

Data documenting the resulting reduction in risk to growers has been rigorously reviewed and acknowledged via the Federal Crop Insurance Corporation's Biotech Yield Endorsement, a risk management instrument that offers an insurance premium rate reduction for farmers using a suite of biotechnology traits.

In 2008, farmers in the United States have planted 11 million ha of triple-stacked corn containing biotechnology traits that provide resistance to corn borers, corn rootworm, and the herbicide glyphosphate. While the yield benefits of these biotechnology traits vary from year to year, this level of planting will increase corn supply in the United States by approximately 5 per cent if yield results seen in 2005 to 2007 are manifested in the 2008 growing season. The contribution of biotechnology traits to world corn supply will increase as they are used more widely.

The next generation of commercialized, biotechnology traits is likely to have a larger impact on crop yields. Improved drought tolerance will be one of the next major, transgenic technologies brought to the marketplace. Drought tolerance has the potential to (1) increase yields in drier areas, (2) increase average yields in rain-fed systems by reducing the effects of sporadic drought, and (3) decrease water requirements in irrigated systems. Similarly, biotechnology traits that improve yield or oil concentration in soybean (Lardizabal et al., 2008) should improve global supplies of vegetable oil and protein meal.

The first of these biotechnology traits, a higher-yielding, glyphosphate-tolerant soybean will be offered commercially in 2009 and commercialization of improved drought tolerance traits is expected around 2012. The transgenes described above are at relatively advanced stages of commercial development. A larger collection of transgenes derived from large-scale screening programmes such as those described by Riechmann et al. (2000), Van Camp (2005), and Creelman et al. (2008) are at earlier stages of development. Biotechnology traits that improve grain yield and nitrogen use efficiency in replicated multiyear field trials are expected to reach farmers' fields in the second half of the next decade.

This group of genetically identified biotechnology traits are referred to as yield-enhancing traits, but the increase in yield may be due to an increase in yield potential and/or an improvement in tolerance to one or more stresses. Collectively, the next generation of biotechnology traits should contribute significantly to productivity on existing cropland, thereby increasing grain supplies and reduce the need to bring new land into production.

While breeders, agronomists, and farmers are working to increase yields, a number of factors that may reduce yields must be considered. Over the next two decades, climate change effects in the central United States are predicted to increase night air temperatures, the number and severity of adverse weather events, and increase the incidence of insect pests and disease. The result could be a drag on crop yields.

Rapid adaptation of crops to changing climatic conditions may help mitigate these effects. Such rapid adaptation may occur for crops supported by strong breeding programmes that continuously develop and introduce superior, locally adapted hybrids and varieties. New, higher-yielding hybrids produced from Monsanto's North American corn breeding programme currently have a product half-life of approximately 4 years and are completely turned over about every 7 years, raising hopes that, for corn and possibly soybean, new varieties adapted to local conditions will be produced as a part of the ongoing breeding programme. Unfortunately, this is not the case for crops such as wheat and rice (*Oryza sativa*) that lack the support of large, private breeding programmes.

Accordingly, increased public support for crop improvement efforts are sorely needed if new wheat and rice varieties are to be adapted to changing local climatic conditions. This is particularly true for regions of the world predicted to undergo more dramatic near term changes in climate than the central United States.

Nitrogen is another factor that may limit crop yields. Nitrogen may become less available as the cost of fertilizer rises and the continued growth of eutrophic dead zones and nitrous oxide emissions leads to changes in the way fertilizer is used. Nitrogen use efficiency, defined as the amount of crop produced per unit of input, has steadily improved in the United States since the 1980s. More precise nitrogen applications and genetic improvements in crops are likely to sustain improvements in nitrogen use efficiency although there is a limit to how far nitrogen application can be reduced. A 10 tonnes/ha corn crop contains around 100 kg nitrogen/ha as protein and at least this amount of nitrogen must be added back to the field to maintain fertility. Lastly, a sharp downturn in the global economy could restrict demand for both meat and fuel in ways that reduce the economic incentive to increase crop yields.

The combination of marker-assisted breeding, biotechnology traits, and continued advances in agronomic practices has the potential to double corn yields in the United States over the next two decades. Doubling U.S. average yields would raise average yields to approximately 20 tonnes/ha, values now seen rarely in non-irrigated corn.

The theoretical light-limited maximum for corn yields in the United States has been estimated at approximately 25 tonnes/ha, close to the 24.2 tonnes/ha recorded on a 2-ha plot by David Hula of Charles City, Virginia in the 2007 National Corn Growers Association Corn Yield Contest. This fact suggests that corn yields can be doubled without large increases in yield potential, although significant improvements in broad stress tolerance, water use efficiency, and broad dissemination of excellent agronomic practices will be required to approach the Hula yield on broad acreage. Improving yields in corn and other crops on a global basis will allow farmers to meet global demand for feed, fuel, and food while minimizing the need to bring large amounts of new land into crop production.

Even if crop producers supported by the agricultural sector fall short of doubling yields, continued public and private investment in agricultural technology will lead to significant increases in productivity that will help supply the world's needs for food, meat, and energy in a sustainable fashion.

6

Increase Phosphorus Use Efficiency to Crop Production

PHOSPHORUS FERTILIZER APPLICATION IN CROP PRODUCTION

Phosphorus (P) is an essential plant nutrient required for optimum crop production. Phosphorus deficiencies can be corrected with phosphate fertilizer (P_20_5). Generally, P is the second most limiting soil nutrient in crop production in Alberta. With respect to fertilizer use, it is second only to nitrogen (N) in Alberta.

EFFECT ON CROP GROWTH

Plants need phosphorus for growth, utilization of sugar and starch, photosynthesis, nucleus formation and cell division. Phosphorus compounds are involved in the transfer and storage of energy within plants. Energy from photosynthesis and the metabolism of carbohydrates is stored in phosphate compounds for later use in growth and reproduction.

Phosphorus is readily translocated within plants, moving from older to younger tissues as the plant forms cells and develops roots, stems and leaves.

Adequate P results in rapid growth and early maturity, which is important in areas where frost is a concern. Frequently, P will enhance the quality of vegetative crop growth.

An adequate supply of available P in soil is associated with increased root growth, which means roots can explore more soil for nutrients and moisture. Phosphorus occurs in most plants in concentrations between 0.1 and 0.4 per cent, on a dry weight basis. A deficiency of P will slow overall plant growth and delay crop maturity.

CONTENT AND CROP REQUIREMENTS

In young, actively growing plants, P is most abundant in the actively growing tissue. By the time plants have attained about 25 per cent of their total dry weight, they may have accumulated as much as 75 per cent of their total phosphorus requirements. Therefore, most crops require significant quantities of P during the early stages of growth. For example, cereal crops will often take up to 75 per cent of their P requirements within 40 days after crop emergence.

Phosphorus requirements for optimum yields vary with different crops. For example, wheat requires less P than canola due to the lower protein content of the seed. A 2,700 kg/ha (40 bu/ac) wheat crop requires about 33 kg/ha (29 lb/ac) of phosphate.

Table. Approximate range of phosphate requirements of wheat, barley, canola and pea

Crop	**Crop part**	**Phosphate kg/ha**	**Phosphate lb/ac**
Wheat 2,690 kg/ha (40 bu/ac)	Seed	23 - 28	21 - 26
	Total Uptake	32 - 38	29 - 35
Barley 3,226 kg/ha (80 bu/ac)	Seed	33 - 40	30 - 37
	Total	44 - 53	40 - 49
Canola 1,960 kg/ha (35 bu/ac)	Seed	36 - 44	33 - 40
	Total	50 - 61	46 - 57
Pea 3360 kg/ha (50 bu/ac)	Seed	34 - 41	31 - 38
	Total	41 - 50	38 - 46

DEFICIENCY SYMPTOMS

A mild P deficiency results in somewhat stunted crop growth, which can be difficult to see. In severe cases of P deficiency, symptoms include characteristic stunting, purpling or browning, appearing first on the lower leaves and base of the stem and working upward on the plant, particularly on cereal crops. The effect is first evident on leaf tips, and then progresses towards the base. Eventually, the leaf tip dies. However, visual diagnosis of P deficiency is very difficult and must be confirmed with soil tests and possibly with the aid of plant tissue analysis.

Symptoms are most pronounced in young plants because their more rapid growth makes greater demands on the available supply. Crops seldom completely outgrow a P deficiency; the symptoms often persist to delay maturity.

SOIL PHOSPHORUS

"Plant-available soil phosphorus" is a term used to mean the portion of soil P that can be taken up from soil by crop roots. It also refers to the portion of soil P extracted by various methods in soil testing laboratories.

In their native state prior to cultivation, Alberta soils often had total soil P levels in the range of 1,100 to 1,350 kg/ha (1,000 - 1,200 lb/ac) in the top 15 cm (6 inches). However, the portion of usable or plant-available P in native soils was very low. Much of the native soil P is contained in soil minerals and in soil organic matter in forms that remain unavailable to plants.

The phosphorus available to plants can be assessed by measuring the phosphate concentration in the soil solution and the soil's ability to maintain the soil solution concentration. The quantity of P in the soil solution, even when at relatively high levels, is only in the range of 0.3 to 3.0 kg/ha (0.3 - 3.0 lb/ac). Rapidly growing crops will absorb about 1 kg/ha (1.0 lb/ac) of P per day.

Therefore, soil solution P must be replenished by the "labile" pool of soil P. Labile P is a pool of soil P that is less available to plants but can undergo rapid chemical or biological changes to recharge or replenish the available P.

Field research in Alberta with a different number of P soil testing methods has shown that the best soil test method for determining plant-available P, called the modified Kelowna method, performs more effectively over a wide range of Alberta soil types.

Remember that soil tests cannot predict with 100 per cent accuracy when crops will respond to added phosphate fertilizer. The frequency of crop response to added phosphate fertilizer can be strongly influenced by environmental conditions, particularly soil temperature and moisture.

For example, at research sites in Alberta with wetter, cooler spring soil conditions, the observed response to phosphate fertilizer, particularly with wheat, barley and canola, tended to be greater than in sites with warmer, drier spring soil conditions. Therefore, producers can expect greater crop response to phosphate fertilizer in a year with wetter and/or cooler spring conditions than in a year with warmer, drier conditions. As a result, soil test recommendations cannot be 100 per cent accurate at predicting when crops will respond to phosphate fertilizer.

It is also important to note that P levels in some soils have increased over the years as a result of repeated annual commercial fertilizer P application or frequent livestock manure application. Consequently, crops grown on some soil types, with higher versus lower soil P levels, are less responsive to fertilizer P application. Additionally, factors such as rate of P fertilizer applied and method of application used can all affect P uptake.

Crop response to applied P fertilizer depends, to a large extent, on the quantity of plant-available P already in the soil. The general soil test ratings for P. The soil test ratings are normally based on a 0 to 15 cm (0 - 6 inch) sample depth because P is not very mobile in the soil. Therefore, the concentration of P is greatest in the surface soil.

Table. Soil test rating for plant-available P levels

Soil test level rating	Phosphorus (P) (lb/ac)
Very low	0 - 20
Low	20 - 35
Medium	35 - 50
High	50 - 80
Very high	greater than 80

Soil P occurs in both organic and inorganic (mineral) forms. Most Alberta soils are relatively low in minerals that contain P, resulting in low plant-available P. Organic P is contained in organic matter and is released slowly by soil micro-organisms.

Micro-organisms

Some soil P is contained in soil micro-organisms. A proportion of inorganic P may be "biologically fixed" by micro-organisms when soil P levels are low. In some cases, micro-organisms may even compete with plants for P when soil P levels are low.

Phosphorus is temporarily tied up in the organic components of micro-organisms; however, this P is eventually returned to the soil when microbes die and break down. After mineralization (conversion from organic P to inorganic P), soil P can be taken up by plants.

Soil pH

Plant availability of P can be affected by soil pH. For example, some P forms are absorbed more readily than other forms. Generally, soil P is slightly more available in a pH range of 6.0 to 7.5 pH.

At higher pH levels (>7.5), calcium may react with phosphorus, creating forms that have slightly lower availability to plants. Magnesium acts in the same manner, forming different types of magnesium phosphate compounds.

In more acidic soils (pH <6.0), iron and aluminum increase, which causes either a fixing or removing of P from the soil solution. This action greatly limits the availability of inorganic P to plants at soil pH levels <5.0.

As a general rule, the maximum availability of P occurs in Alberta soils within a pH range of 6.0 to 7.5.

Soil moisture and temperature

In general, soil moisture and temperature affect P availability and root growth. In cool, wet soils, P availability and movement are reduced. As a result, crops are more responsive to phosphate fertilizer in cool, wet spring conditions than in warmer, drier spring conditions.

Optimal soil moisture and temperature can help accelerate microbe activity, thereby releasing more P from organic matter. Adequate soil moisture will enhance fertilizer solution and reaction in the soil. As well, moisture will promote plant growth, so P and other nutrient requirements are generally higher for crops grown under irrigation or in higher rainfall areas.

COMMON PHOSPHORUS FERTILIZER TYPES

Two of the more common phosphate fertilizers sold.

Table. Common phosphate fertilizer types sold in Alberta

Name	Type	Form
Mono-ammonium phosphate	12-51-0	Granular
Ammonium polyphosphate	10-34-0	Liquid

Mono-ammonium phosphate (ortho phosphate), which is the most common form of phosphate used in Alberta, can be safely blended with urea or ESN to produce various blends. It can also be blended with muriate of potash (0-0-60; 0-0-62) or with most sulphur fertilizers as well.

Most soil testing laboratories in Alberta report soil test P level in lb/ac, but fertilizer recommendations are made in terms of phosphate (P_20_5) in lb/ac. To convert P to P_20_5, multiply by 2.3 (*e.g.* 10 lb/ac of P = 23 lb/ac of P_20_5). Phosphorus fertilizer in Canada is sold on the basis of P_20_5 in per cent of total product.

PHOSPHATE FERTILIZER APPLICATION

There are a number of methods of applying P fertilizer. The fertilizer placement; however, some methods are more efficient than others. To obtain good P fertilizer efficiency, adequate rates of nitrogen and other nutrients must be available to the crop.

Broadcast and incorporation

The broadcast and incorporation method involves uniformly broadcasting and incorporating P fertilizer onto the soil before seeding. At low and medium soil test levels for P, the broadcast and incorporation method is much less effective than seed-placed or banded P.

When P is broadcast and incorporated, application rates may have to be two to four times the in-row banded rate to obtain an equal crop response when soil test P level is very low. Therefore, broadcast and incorporation of P may not be economical or practical if high rates are needed to increase yield.

However, high rates of broadcast and incorporated P on eroded soil areas or knolls are recommended to even out the fertility of eroded fields because these fields are often low in soil P. Eroded soils are often very responsive to P fertilizer.

Broadcasting P at higher rates involves a high initial cost, but these costs can be recovered over several years. The availability of applied P to plants tends to decrease over a period of years because of the reaction products formed in the soil. However, it is also recommended that the one-time large application be followed by an annual application of seed-placed P fertilizer with annual crops.

When high rates of P fertilizer are applied, it is possible to reduce the availability of micronutrients, particularly zinc (Zn) and copper (Cu). In southern Alberta, where high rates of P have been applied, isolated cases of zinc deficiency have been observed during cool, wet spring conditions in both irrigated field beans and in corn grown on sandy soils where high P rates have been used. In central Alberta, copper deficiency in cereal crops has been associated with higher phosphate fertilizer applications.

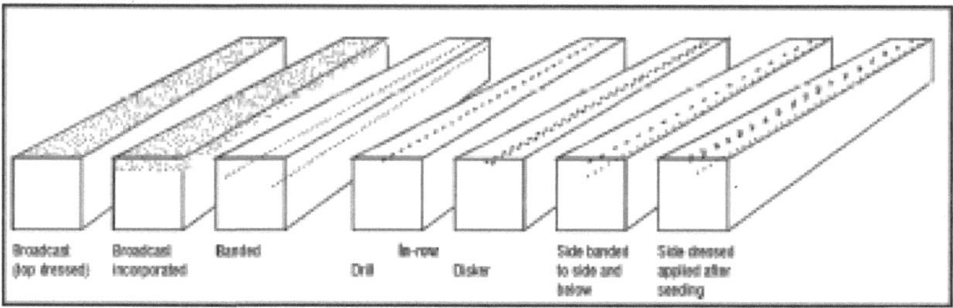

Fig. Various methods of phosphate fertilizer application.

Top dress (broadcast - no incorporation)

Broadcasting P fertilizer without incorporation is only recommended for established forage crops because it is the only practical application method currently available. Granular phosphate can be broadcast. Liquid phosphate fertilizer (*e.g.* 10-34-0) can be dribble banded onto forage stands. There are times where dribble-banded liquid P fertilizer might be superior to a broadcast application of granular P fertilizer. However, liquid fertilizer is generally more expensive per pound of phosphate versus granular fertilizer.

Phosphorus fertilizer is immobile in soil; therefore, plant uptake of fertilizer P may be low in the first year after application. However, alfalfa and grasses do have feeder roots very near the soil surface and can take up some broadcast P fertilizer when surface soil moisture conditions are good.

For forage crops, a three- to four-year supply of P can be either deep banded or broadcast and incorporated before establishment. Subsequent applications may be top dressed. Recent work in southern Alberta showed that annual broadcast P application was as good as and at times superior to a one-time large application of P fertilizer prior to establishment of timothy hay.

Seed-placed phosphorus (banded with seed in the seed row)

The maximum safe rate of P that can be applied with the seed for cereal crops with a 10 per cent seedbed utilization (SBU) is 50 to 70 kg/ha (45 - 65 lb/ ac) of P_20_5depending on soil moisture conditions and the opener used. For oilseed crops, seed-placed P_20_5 rates should not exceed 15 kg/ha (14 lb/ac) when a seeding implement that places the seed and fertilizer in a narrow band (10 per cent SBU) is used. For peas, seed-placed P_20_5 rates should not exceed 30 to 35 kg/ha (27 - 33 lb/ac), especially when a seeding implement that places the seed and fertilizer in a narrow band (double disc drill) is used.

Single and double disc no-till drills cut a fairly narrow furrow and place the seed and fertilizer together in the bottom of the furrow. These drills place the fertilizer in a concentrated band close to the seed and are most likely to result in damage to sensitive crops at higher rates of application.

Hoe-type openers cut a wider furrow, and the seed and fertilizer are scattered across the bottom of the furrow. This scattering in a wider band results in a lower concentration of fertilizer close to the seed and is less likely to cause damage to sensitive crops such as canola or flax when high rates of fertilizer are applied.

Seeding implements that scatter the seed in wide bands (air seeders with sweeps) reduce the seed-fertilizer contact. As a result, higher seed-placed P rates can safely be used.

However, because of the wide variation in furrow opener design and spacing, as well as the width of the seed row, specific maximum safe seed-placed P rates have not been determined for all opener types. Generally, rates can be increased by 25 to 50 per cent over the safe rates for a 10 per cent SBU. At low rates of fertilizer application, response to seed-placed P may be slightly less when the seed and fertilizer are spread out in broad bands as compared to narrower bands.

Side-banding phosphorus

The side-banding method places the P in a band near the seed row during the seeding operation. The fertilizer is commonly banded 2.5 to 5 cm (1 - 2 inches) below and/or to the side of the seed row for small seeded crops and row crops such as sugar beets, potatoes, sunflowers, corn and beans. This method is the best placement option for crops that are sensitive to seed-placed phosphate fertilizer.

Phosphate at all recommended rates can be safely applied with the side-banding method. However, specialized side-banding seeding equipment is required. Yield increases obtained with this placement are nearly equal to seed-placement at low rates of application, but much superior when application rates are too high for safe placement with the seed. Phosphorus side banded near the seed of cereal crops in most cases is close in efficiency to seed-placed phosphorus.

Some no-till seed drills use a mid-row or paired-row technique. Two rows are seeded about 7.5 to 12 cm (3 - 5 inches) apart, and P is banded between the two rows about 2.5 to 5 cm (1 - 2 inches) below the seeding depth. There is an 18 to 25 cm (7 - 10 inch) space between the next pair of rows. Generally, this method of placement is effective. However, root access to the P fertilizer could be compromised in the following situation: if the fertilizer is 8 cm (3 inches) more to the side and not below the depth of seeding, seeding depth is relatively shallow and soil conditions dry out after seeding.

Banding phosphorus

The banding method of application places P in concentrated bands in the fall or spring before seeding. A cultivator-type implement with shovels or knives is used with a shank spacing of 20 to 35 cm (8 - 14 inches). Depth of application is normally 8 to 10 cm (3 - 4 inches).

Banded P can be somewhat less effective than seed-placed P when banded with N fertilizer just before seeding. However, both seed-placed and banded P are more effective than broadcast and incorporated P when soil test levels are low.

In some locations and years, particularly on very low P soils and when soil temperature is low following seeding, the "starter" effect of phosphate placed with or near the seed can be important and cannot be achieved by banded phosphate alone. For this reason, seed placement of a portion of the P at 11 to 17 kg/ha (10 - 15 lb/ac) P_20_5 is advised when phosphate is banded.

Banded P may be more effective than seed-placed P under dry surface soil conditions. The major advantages of banding P are as follows:

- Rates considered too high for seed-placement with oilseeds and pulse crops can be used where low soil P warrants such rates
- The amount of fertilizer handled at seeding time can be reduced or, in some situations, eliminated
- The disadvantage of banding P in the spring is that it may dry out the seedbed, which could reduce germination and limit crop yield under dry spring moisture conditions. This concern is particularly important in southern Alberta in drier spring conditions.

Application of phosphate both with the seed and deep banded (split application) is an effective way to achieve the benefits of both methods.

Narrow spacing between phosphate bands is preferred so that phosphate uptake is not delayed during early growth owing to long distances between the bands and some of the seedlings. On very low P soils, "stripping" can occur, which is caused by poorer growth of plants midway between phosphate bands when the bands are too widely spaced. For phosphate, banded spacings of more than 30 cm (12 inches) are not recommended, and narrower spacings may be more effective, particularly where no phosphate is seed placed.

When banding phosphate, the bands must be sufficiently deep to avoid disruption by subsequent tillage and seeding operations. Time of phosphate banding (fall versus spring) appears to have little effect on fertilizer efficiency.

Plant uptake of phosphate placed in the same band with nitrogen (N) can be restricted. For example, Alberta research has shown that when 60 kg/ha (54 lb/ac) of phosphate was banded with 120 kg/ha (108 lb/ac) of N, phosphate uptake efficiency was only 12 to 15 per cent compared to 30 to 35 per cent when the same rate was placed in a separate band away from the N band.

Generally, phosphate should not be banded with N fertilizer if the N rate is higher than 70 to 80 kg/ha (63 - 72 lb/ac) to avoid reduced uptake efficiency of the P fertilizer. The major reason for this effect is that plant roots cannot penetrate the concentrated nitrogen band and, therefore, cannot take up the P effectively.

Some of the advantages of banding before planting include the following:

- Application of fertilizer during the busy seeding period is avoided. However, some growers may still wisely place some P fertilizer with the seed for starter effects.
- Fertilizer nutrients are strategically placed at depths in the root zone where soils tend to remain moist longer during the early parts of the growing season and where they are beyond the reach of shallow germinating and rooting weeds.

Comparison of placement methods

Placing P fertilizers in bands minimizes the contact between the soil and the P fertilizer. In contrast, mixing the P with soil exposes it to more soil, resulting in reduced plant availability. Alberta research suggests that placement of P with the seed is frequently better than banded P. The wheat, barley and canola response to seed-placed versus banded phosphate fertilizer in southern Alberta in 1991 on summerfallow and stubble.

Table. Response to seed-placed versus banded phosphate fertilizer

	Wheat		Barley		Canola	
	Fallow	Stubble	Fallow	Stubble	Fallow	Stubble
Number of sites	7	17	7	19	7	15
P responsive sites	6	10	6	13	6	14
Seed placed > banded	4	6	4	8	5	6
Banded > seed placed	2	3	1	0	0	2
Seed placed = banded	0	1	1	5	1	7

Seed placement of 17 kg/ha (15 lb/ac) is advised when phosphate is banded to obtain a starter effect when soil P levels are very low, 0 to 20 kg/ha (0 - 18 lb/ac) or cool soil temperatures occur after seeding, conditions common in the Black, Gray-Black and Gray soil zones.

Seed-placed P is recommended as it is one of the most efficient means of P application, provided the amounts applied do not injure the germinating seed and seedling. When higher rates of P are used in dry and/or coarse-textured soils, banding away from the seed at planting at times may be superior to placing phosphate with the seed.

Cereal Crops

Soils low and medium in available P

On most soils low in available P, seed-placed phosphate at recommended rates is equal to or better than banding near the seed and far superior to surface application and incorporation. Recommended phosphate fertilizer should be applied, and the probability of crop response would be in the range of 60 to 100 per cent.

If P is broadcast and incorporated, the annual application rate must be two to four times the rate recommended for seed-placement to obtain equal crop response in the year of application at very low and low soil test P levels.

Soils high in available P

In soils high in available P, seed-placed or banded fertilizer P at rates up to 20 to 30 kg/ha (18 - 27 lb/ac) of P_20_5 may result in a crop response 30 to 50 per cent of the time, depending on crop, soil zone and environmental conditions.

Canola and Mustard

Soils low and medium in available P

On soils low and medium in available P, rates up to 15 kg/ha (14 lb/ac) P_20_5 can be seed placed using a seedbed utilization of 10 per cent. Rates greater than 15 kg/ha (14 lb/ac) P_20_5 should be either banded prior to seeding or side banded at the time of seeding.

Soils high in available P

Seed-placed or banded fertilizer P on soils high in available P at rates up to 15 kg/ha (14 lb/ac) of P_20_5 may result in a crop response 30 to 50 per cent of the time, depending on soil zone and environmental conditions.

Table. Phosphate fertilizer recommendations for canola and mustard on a medium to fine textured soil with a neutral pH, based on the Kelowna soil test method. Recommendations are given for each soil zone at three soil moisture condition levels at the time of seeding

Soil test P (lb/ac)	Brown			Dark Brown			Thin Black			Black			Gray Wooded			Irrigated
	D*	M*	W*	D	M	W	D	M	W	D	M	W	D	M	W	
	P_2O_5 (lb/ac)															
0 -10	30	35	40	35	40	45	40	45	50	40	45	50	40	45	50	50
10 - 20	25	30	35	30	35	40	35	40	45	35	40	45	35	40	45	45
20 - 30	20	25	30	25	30	35	30	35	40	30	35	40	30	35	40	40
30 - 40	15	20	25	20	25	30	25	30	35	25	30	35	25	30	35	35
40 - 50	15	15	20	20	20	25	25	25	30	25	25	30	25	25	30	35
50 - 60	15	15	20	15	15	25	20	20	30	20	20	30	20	20	30	30
60 - 70	15	15	15	15	15	20	15	20	25	15	15	25	15	15	25	25
70 - 80	0	15	15	0	15	15	0	15	20	0	15	20	0	15	20	20
80 - 90	0	0	15	0	0	15	0	0	15	0	0	15	0	0	15	15
>90	0	0	0	0	0	0	0	0	0	0	0	0	0	0	0	0

Note:

* Seedbed soil moisture conditions at seeding D = 25 per cent; M = 50 per cent; W = 75 per cent of field capacity.

Table. Approximate probability of a greater than 2 bu/ac and 5 bu/ac canola response to phosphate fertilizer when following recommendations

Soil test P (lb/ac)	Brown		Dark Brown		Thin Black		Black		Gray Wooded		Irrigated
	>2	>5	>2	>5	>2	>5	>2	>5	>2	>5	
						%					
0-10	90	70	95	80	95	90	95	95	95	90	95
10 - 20	80	60	90	80	95	85	95	90	95	85	90
20 - 30	70	60	80	70	90	75	90	80	90	75	80
30 - 40	60	50	70	60	80	65	85	70	80	65	70
40 - 50	50	40	60	50	70	55	80	60	70	55	60
50 - 60	40	30	50	40	60	45	70	50	60	45	50
60 - 70	40	30	40	30	50	35	60	40	50	35	40
70 - 80	35	20	35	20	40	30	50	30	40	30	30
80 - 100	30	10	30	15	30	20	40	20	30	20	20

Pea and Lentil

Phosphorus soil tests and fertilizer recommendations developed for pea and lentil in Alberta are fairly effective in predicting when these crops will respond to P fertilizer application. Soils that have accumulated fertilizer P over the years may still test deficient in plant-available P, particularly on high pH calcareous soils. Yet, crops grown on this land may not respond to added P fertilizer. This situation may occur in the Brown and Dark Brown soil areas of southern Alberta.

The phosphate fertilizer recommendations for pea and lentil in Alberta. P fertilizer does not have a strong effect on pulse crops as these crops are fairly efficient at taking up soil P. Alberta research suggests that pea is most responsive to P fertilizer when soil P levels are less than 30 lb P/ac. Above this level, there is relatively low chance P fertilizer will increase yield. When soil test P levels are medium and significant P fertilizer was applied in the past 10 to 20 years, an annual maintenance application of phosphate fertilizer can be used to meet crop requirements and replenish soil P that is removed.

Phosphate fertilizer applied when soil test P is higher than 30 lb P/ac will help to replace soil P and maintain good soil P levels.

Flax

Flax does not respond as readily to fertilizer P banded with the seed. Also, flax is sensitive to seed-placed P.

Soils low in available P

On soils low in available P, the P fertilizer can be seed placed but should not exceed 15 kg/ha (14 lb/ac) P_20_5, and remaining P should be side banded at seeding or banded prior to seeding.

Soils medium to high in available P

In soils medium to high in available P, banding P prior to seeding or side banding at seeding at rates of 28 kg/ha (25 lb/ac) P_20_5 may result in a crop response, particularly in cool, wet spring conditions.

Corn, Potatoes and Sunflowers

Soils low to medium in available P

On soils low to medium in available P, all P should be banded or side banded at the time of seeding away from the seed row.

Table. General phosphate fertilizer recommendations for pea and lentil grown in the various soil zones of Alberta at increasing levels of soil P, based on the Kelowna method of extraction of soil P

Soil test P (lb/ac)	Brown and Dark Brown	Thin Black and Black	Gray Wooded	Irrigated
	P_2O_5 (lb/ac)			
0 - 10	35	40	35	45
10 - 20	30	35	30	40
20 - 30	25	30	25	35
30 - 40	20	25	20	30
40 - 50	15	20	15	25
50 - 60	15	15	15	20
60 - 70	0	15	15	15
>70	0	0	0	0

Soils high in available P

Banding or side-banding P on soils high in available P may result in a crop response 40 per cent of the time.

Sugar Beets

The general phosphate fertilizer recommendations for sugar beets.

Soils low to medium in available P

On soils low to medium in available P, up to 22 kg/ha (20 lb/ac) P_20_5 can be seed placed.

Rates greater than 22 kg/ha (20 lb/ac) P_20_5 should be banded prior to seeding or side banded near the seed.

Soils high in available P

On soils high in available P, banding or side-banding P with or near the seed in these soils may increase seedling vigour but may or may not carry through to increase yield.

Forage Crops

The best method and time of fertilizer P application for cultivated grasses, grass-legumes and legumes are broadcast-incorporated or deep banded prior to stand establishment coupled with annual broadcast application as needed. Phosphorus applied in this manner will suffice for the year of establishment and for two to four years after, depending on the initial rate of application, available soil P levels and other soil characteristics.

On soils low in available P, a single high rate (100 to 150 kg/ha or 90 to 135 lb/ac) of P_20_5 broadcast and incorporated prior to seeding has shown better yield response over a three-year period than an equivalent amount broadcast in annual increments in some situations.

However, P requirements for established forage stands can be supplied by broadcast applications, particularly under irrigated conditions and for wetter areas of the province. In the future, it may be possible to band granular or liquid fertilizers into established forage fields by using very narrow openers or by injecting liquid fertilizer using extremely high pressure. However, this technology has not been developed yet.

Table. General phosphate fertilizer recommendations for crops grown in Alberta at increasing levels of P in the soil based on the Kelowna method of extraction of soil P

Soil test P (lb/ac)	Flax	Potatoes	Sugar beets	Corn and sunflowers
	P_2O_5 (lb/ac)			
0	35	100	80	70
10	30	90	70	60
20	25	80	60	50
30	20	70	50	40
40	15	60	40	30
50	15	50	30	20
60	15	40	30	20
70	15	40	30	20
80-100	10	30	20	15
>100	0	0	0	0

IMPROVING PHOSPHORUS EFFICIENCY IN CEREAL CROPS

The majority of the world's mined rock phosphate is used for the manufacture of phosphorus (P) fertilizers applied in agriculture to improve or sustain crop yields. Over 17.5 mt P per year was used for agricultural purposes from 2004 to 2008, and this figure is projected to increase over the coming years. While P fertilizer prices remained at a stable and low level for several decades, they have more than doubled recently and considering the non-renewable nature of rock-P resources, further price increases seem inevitable.

Increasing the efficiency with which P is used in agricultural systems is therefore critical for sustainable food and fibre production in the twenty-first century.

Crop P efficiency (PE) can improve if either yields increase at a given rate of P fertilizer application, or if yields remain stable with lower levels of P fertilizer application. This could be broadly achieved by enhancing P uptake (P acquisition efficiency; PAE) or by improving internal P utilization efficiency (PUE). Given that only a portion of the P applied in fertilizers is actually taken up by a crop on high P fixing soils, with the remaining P being slowly immobilized in the soil, there seems to be opportunity for further improvement in PAE. Several reviews on the topic of enhancing PAE have recently been published and the reader is referred to Richardson et al. (2009) for further discussion.

Far less research has been conducted on the topic of how to improve PUE. While numerous agronomic definitions of PUE exist (*e.g.*, grain yield per unit of P fertilizer applied, grain yield per unit of P in aboveground biomass), from a physiological and breeding perspective we define PUE as shoot biomass produced per unit P in shoots. Essentially, plants with higher PUE operate at lower shoot P concentrations. During the vegetative growth phase most plant P is contained in shoot tissue and shoot P concentrations (as opposed to root P concentrations) are therefore of primary importance for improved PUE. However, at harvest, cereals typically contain 70^{+} per cent of their total P in grains with very little remaining in straw, prompting us to investigate whether reductions in grain P concentrations are a possible way to improve overall crop PE in cereal systems. Previous studies have already demonstrated that grain P can be lowered through recurrent selection or by mutation. In this opinion piece, we discuss the advantages and possible disadvantages of a low grain P trait in the context of breeding P-efficient crops for sustainable cropping systems.

REDUCING P FLOWS THOUGH CROPPING SYSTEMS: IMPLICATIONS FOR FERTILIZER REQUIREMENTS AND THE ENVIRONMENT

The importance of P in harvested products is highlighted by calculations showing that the total P removed annually in grain and fleshy fruit crops equates to 85 per cent of fertilizer P applied to crops globally. While this calculation would suggest that the global P balance is positive, it has to be kept in mind that huge imbalances in the application of P fertilizer exist across the globe, with P surpluses in many European and East-Asian countries and large P deficits in many poorer areas including sub-Saharan Africa. The high P removal rate in harvested grains drives the need to replace soil-P by fertilizer application or leads to P mining where fertilizer application rates are low. Developing crop varieties that translocate less P to developing grains may offer one option to balance P budgets in agriculture at reduced fertilizer requirements.

Low grain P crop varieties may also have environmental benefits because much of the P removed from fields ultimately ends up in landfill or water bodies via sewage pathways. This occurs because there are losses of P through inefficiencies at each step in the food chain, including inefficient recovery or suboptimal redistribution of high-P animal wastes, losses of P during food manufacturing processes, and particularly poor recovery and use of P in human waste.

One of the major causes of inefficiency is phytate, the organic storage form of P which usually accounts for around 75 per cent of total P in grains, because animals – particularly monogastric animals including humans – cannot properly digest phytate, leading to high P concentrations in feces and urine. In the case of livestock, the poor digestibility of phytate can also lead to P deficiency in the animals, which has traditionally been rectified by using P feed supplements or by the addition of phytase enzymes to feed rations. More recently, low phytic acid (*lpa*) mutants have been identified in major crop species, which typically have a reduced concentration of phytate with concurrent increases in inorganic P in grains, providing better P nutrition in livestock and a reduction of P in feces and urine.

In contrast to livestock, humans rarely suffer from P deficiency, so additional inorganic P in grains for human consumption is of little benefit and P concentrations in human waste will remain high. While recycling of P in human and animal waste needs to be improved globally to alleviate environmental effects and close the P loop, it seems unlikely that it will occur rapidly and efficiently enough over the coming decades to prevent substantial loss of valuable P from agricultural systems. Thus, attempting to minimize the throughput of P in the agricultural sector would be the best way of minimizing negative environmental impacts while maintaining high productivity. The most logical way to reduce the throughput is to address the removal of P from farms in harvested grain.

Interestingly, the concept of reducing grain P has been raised on numerous occasions in the past but there has not yet been a global effort to address the issue, presumably because of the association between seed P levels and seedling vigour. We argue that given the economic and environmental gains to be made with a low grain P trait, seedling vigour should not be seen as an impediment for further exploration, but rather an issue that should also be addressed through further research.

PHOSPHORUS EFFECTS ON SEED GERMINATION AND SEEDLING VIGOUR

Early crop vigour is critical in productive farming systems and typically relates to higher biomass and/or grain yields. Sustaining or increasing seed P content is thought to improve seedling establishment and vigour. Two separate lines of enquiry have contributed to these conclusions: the first of these was a series of studies from the 1980s and 1990s which reported that seeds obtained from plants grown in P-deficient soil (which had lower P concentrations than seeds obtained from plants grown in high-P soils) had poor germination and establishment, and subsequent lower biomass or grain yields.

We recently conducted a study using low-P rice seed obtained from plants grown in P-deficient soil and found that seed germination and seedling vigour were impaired, as per the earlier studies mentioned above. However, we concluded that a significant proportion of the reduced germination and seedling vigour was likely due to the severe P deficiency stress suffered by the mother plants during grain filling, rather than caused by low seed P concentration *per se*. Further experiments using seeds differing in P concentration that were not obtained from plants grown under severe P deficiency found no differences in yields when plants were grown in P-replete or P-deficient soil.

More recently, experiments have suggested that genotypic differences have a much larger effect on seedling vigour than seed P concentrations or total seed P content. Even seed P concentrations as low as 1 mg g^{-1} did not impair germination, seeding vigour, or final grain yield in a generally vigorous rice genotype, presumably because such vigorous genotypes rapidly compensate for low seed P by seedling P uptake (Wissuwa, unpublished data).

A second line of enquiry using *lpa* mutants or transgenic plants with altered phosphatase scavenging capacity has also suggested that alterations to seed P affect seed germination and seedling vigour. Studies with *lpa* mutants across a range of species found reduced seedling vigour or impaired germination compared to wild-type parent lines.

However, most *lpa* mutations impair the function of synthase or kinase enzymes involved in inositol phosphate metabolism, and this metabolism is important in vegetative tissues as well as developing seeds. Hence, perturbation of the phytic acid synthesis pathway can have downstream effects on seed viability, germination and seedling vigour, and general plant growth, especially under stress conditions. Similarly,Robinson et al. (2012) recently reported that eliminating the purple acid phosphatase, AtPAP26, in*Arabidopsis thaliana* decreased seed P levels but also reduced seed germination. These authors concluded that reducing P translocation to the seed is undesirable; however, the poor germination could also be explained by the fact that AtPAP26 probably has a critical role in the seed germination process.

In all of the above-mentioned cases, it is plausible that the disruption of key metabolic pathways induced by mutation leads to problems with seed germination and seedling vigour rather than a deficiency of P in seeds*per se*. Indeed, proof-of-concept that a reduction in seed total P does not necessarily impact on subsequent crop performance is provided by the barley *lpa1-1* mutant. This mutant has a 10–20 per cent reduction in seed P, and barley isolines with the mutation do not suffer from an obvious yield penalty across a range of environments. Importantly, the mutation – to a gene which appears to encode for a member of the sulfate transporter gene family – appears to have a seed-specific impact by reducing endosperm P levels by 30 per cent without broader negative effects on plant physiological processes. The fact that commercial barley cultivars have been developed and released in North America suggests that reducing the P concentration of cereal crops by around 20 per cent through breeding without reducing subsequent crop yields is certainly feasible.

THE NEED FOR A LOW GRAIN P TRAIT TO IMPROVE INTERNAL PUE

Internal PUE has received significant attention recently due to recognition that enhanced PAE alone cannot improve the PE of farming systems. However, the target of improving PUE, *i.e.*, increasing biomass production per unit P, may clash with the perceived need to maintain high grain P concentrations. This issue is rather elegantly explained by Barraclough et al. (2010)with regard to internal nitrogen use efficiency (NUE) and grain protein concentrations, whereby high NUE is likely to come at the cost of reducing grain protein. These authors applied what is known as the "law of conservation of matter" to NUE, and the concept applies to any nutrient, including P. Reduced shoot P concentrations (higher PUE) will logically lead to reduced grain P concentrations unless some way can be found to uncouple both traits. Such uncoupling could theoretically be achieved by increasing the proportion of P located in grains at harvest. However, given that the PHI of most crops is already general above 70 per cent and can be up to 90 per cent, scope to enhance this further appears rather limited.

We have simulated PUE improvements in rice permissible under the "law of conservation of matter" based on different grain P and PHI scenarios. Calculations suggest that if grain P concentrations were sustained at a given level (*e.g.*, 2.5 mg g^{-1}), an increase in PHI from 0.7 to 0.8 would allow for moderate reductions in shoot P concentration at flowering – *i.e.*, an increase in vegetative PUE.

However, much larger gains in PUE at flowering are only possible if grain P concentrations are reduced.Rose et al. (2010) suggested a breeding target of a 20 per cent reduction in grain P in rice (*e.g.*, from 2.5 to 2.0 mg g^{-1}), which, using the rough calculations, would allow for a reduction in shoot P concentration at flowering of between 0.4 and 0.7 mg g^{-1}, depending on the PHI and amount of P taken up from the soil during grain filling.

While shoot P concentrations of most cereal crops range from around 4 mg g^{-1} at tillering to around 2 mg g^{-1} at flowering, as little as 1 mg g^{-1} P may actually be necessary for normal cellular function. This is consistent with reports that reduced growth under P deficiency may be due to gene expression reprogramming rather than a direct result of low levels of shoot P. Given the fact that it may be possible to repress the gene expression reprogramming observed under P deficiency it appears that there is a real prospect of improving PUE in crop plants. However, any reduction in shoot P concentrations towards the theoretical limit of around 1 mg g^{-1} will not be possible without a subsequent reduction in grain P concentration.

P EFFICIENCY VERSUS P BALANCE EFFECTS

Grain P concentrations could therefore, in theory, be reduced passively as the result of improvements to crop PUE or actively by breeding specifically for a low grain P trait. If a reduction in grain P concentrations is the result of improvements to PUE, one could expect that there would be a decline in P fertilizer requirements (because presumably the high PUE plants would not need to acquire as much soil P as current cultivars), as well as a "slowdown" in global P cycling as the throughput of P is minimized.

The reduced throughput of P in the system would be driven by the lower P fertilizer requirements of crops but the low grain P trait would also minimize the subsequent P flow through the food chain. However, in reality breeding for improvements to PUE may be challenging in high-input farming systems. In the absence of enhanced PUE, breeding directly for a low grain P trait would still reduce P flow through the food chain and minimize soil P mining, but whether the return of high-P straw residue to the soil could reduce subsequent P fertilizer requirements is not known.

One question that must be addressed is whether additional P retained in straw would be bioavailable or whether it would simply contribute to the build-up of recalcitrant organic P in soils. Given that contamination of waterways with P from diffuse sources (*e.g.*, agricultural run-off) is the major contributor to eutrophication, the continued build-up of P in soils is a major concern. Recalcitrant P in soils is a particular problem because it is of little benefit to plants, but can become biologically active while residing in the sediment layer at the bottom of water bodies.

However, the general perception that the bulk of P returned to soil in organic material is in an organic form is largely unfounded and much of it is present as inorganic P or readily degradable organic P forms. Although inorganic P from crop residues can become immobilized in soil organic P through microbial processes, the same processes also apply to P applied as inorganic or organic fertilizer.

Ultimately, the value of P returned in crop residues may depend on breeding crops that can utilize these P sources directly or indirectly through microbial mediated pathways.

GRAIN P IMPLICATIONS FOR HUMAN HEALTH

Phosphorus is the second most abundant mineral in the human body after calcium, and 85 per cent of it exists in bone. Dietary P intake is critical because both P deficiency and excess can damage bone health. The average recommended P intake is 700 mg/day, varying from 100 mg/day infants to 1250 mg/day for 9 to 18-year-olds. The greatest contributors to P intake are protein-rich foods (*e.g.*, dairy products) and cereal grains. The P intake from cereals alone can reach 400–600 mg/day in some Asian countries such as Japan where cereals (*e.g.*, rice) are a staple food. Therefore, reducing P concentrations or modifying the composition of P-containing compounds in grains may have significant effects on human health which warrants careful consideration.

Phosphorus in cereal grains is present in different chemical forms including inorganic phosphate, phytate, phospholipids, DNA, RNA, and ATP, though phytate and phospholipids have the most impact on human health. While phytate is poorly digestible and regarded as an anti-nutrient because it reduces the bioavailablity of micronutrients and protein, it has been reported to have positive effects – it may protect against cancer and may help treat diabetes mellitus, atherosclerosis and heart disease.

However, these authors note that the effective dosage to elicit these beneficial effects is still not clear. Compared to phytate, cereal phospholipids provide a better source of bioavailable P and vitamins such as choline. Dietary phospholipids have beneficial effects on many human diseases including coronary heart disease, cancers, and inflammation. Cereal starch also contains a significant portion of lysophospholipids which naturally form an inclusion complex with amylose. When the lysophospholipids form a complex with amylose, they may slow down the rate of digestion and adsorption of starch which may be beneficial to diabetics.

While there has been a substantial effort to breed or engineer crops with reduced phytate but unchanged total P in grains, the impact of such crops on human health are still unknown. As phospholipids have clearer nutritional values and health benefits than either phyate or inorganic phosphorus, there may be scope to breed healthier cereal grains by lowering phytate but increasing phospholipids contents, in particular the starch lysophospholipids which are located in the endosperm (*i.e.*, in white rice, which is most commonly consumed). Research into grain phospholipids and the potential to maintain or increase the concentration of starch phospholipids while reducing phytate is ongoing.

PHOSPHORUS FERTILIZER USE EFFICIENCY FOR FLORIDA'S HORTICULTURAL CROPS

WATER-SOLUBLE N AND P FERTILIZERS

Water-soluble fertilizers dominate in Florida because they are plentiful and cost considerably less than slow or controlled-release fertilizers. A negative aspect about water-soluble N fertilizer is that it readily and rapidly leaches through Florida's sandy soils with excessive rainfall or irrigation. Five materials make up essentially the entire suite of water-soluble N fertilizers used. Each material has unique characteristics that make it suitable or unsuitable for particular horticultural applications. For example, the acidifying property of ammonium sulfate makes it desirable for application to acid-loving plants such as highbush blueberry (*Vaccinium corymbosum*). On the other hand, urea is not suitable for application to Florida's alkaline soils due to ammonia volatilization hazard.

Phosphorus fertilizers applied in Florida are almost entirely water-soluble materials. An exception occurs when a polymer-coated N–P–K material is used that applies a small amount of P in controlled-release form. The group of water-soluble P fertilizers used in Florida is comprised of four materials that each have unique characteristics. They are quite versatile in their horticultural application, but there are a few instances where a particular material should not be used. For example, diammonium phosphate should not be applied to an alkaline soil due to volatilization of N and loss of P availability.

The efficiency of WSFs can be enhanced by using appropriate rates, sources, application timing, and placement. Specific techniques to improve WSF efficiency in Florida's horticultural production systems are discussed in production guides.

BMPS TO IMPROVE WSF USE EFFICIENCY

In Florida, BMPs are defined as "practices based on research, field-testing, and expert review determined to be the most effective and practicable on-location means, including economic and technological considerations, for improving water quality in agricultural and urban discharges." The formal procedure used to establish horticultural crop BMPs after the need was identified included: 1) forming a steering committee and technical work groups, 2) producing a draft BMP manual, 3) peer-review and revision, 4) holding public workshops, and 5) adopting the manual in state code.

Producers who voluntarily implement BMPs receive a presumption of compliance with water quality standards and are eligible for cost-share funds to improve operations.

Florida Department of Agriculture and Consumer Services (FDACS) has published BMP manuals for sod, container nursery, vegetable, and citrus production, while the Florida Department of Environmental Protection (FDEP) has produced BMP manuals for golf courses and turf/ornamental landscapes. Most nutrient BMPs are simple, common-sense, "good housekeeping" practices. In abbreviated form, they involve: 1) educating and training field operators about how to manage fertilizer, 2) developing a nutrient management plan, 3) using appropriate fertilizer sources and formulations, 4) using application equipment that applies fertilizer to target sites, 5) properly calibrating and maintaining application equipment, 6) using precision nutrient application where appropriate, 7) avoiding high-risk fertilizer applications (such as during the rainy season), 8) splitting fertilizer applications throughout the growing season, 9) adding organic matter to the soil whenever possible, and 10) using soil moisture sensors to improve irrigation scheduling.

ENHANCED EFFICIENCY FERTILIZERS (EEFS)

The American Association of Plant Food Control Officials describes EEFs as fertilizer products with characteristics that minimize the potential of nutrient losses to the environment, as compared with "reference soluble" products. Enhanced efficiency fertilizers include slow and controlled-release materials as well as nitrification and urease inhibitors.

Agronomic effectiveness of EEFs is not implied in this definition, but it is assumed. In Florida, high-volume use of EEFs in horticultural applications is limited to turfgrass, greenhouse, nursery, and landscape settings due to higher costs compared with WSF. Smaller amounts of EEFs are used in citrus replant situations and other specialty horticulture. For example, vegetable growers sometimes include slow-release N in fertilizer blends applied under plastic mulch.

Nutrient release from EEFs

There are several modes by which nutrients are released from EEFs. Coated fertilizers contain water-soluble N (alone or in combination with other nutrients) surrounded by an impermeable or semipermeable coating. Nutrients are released by diffusion through the coating or following degradation of the coating.

Examples include sulfur-coated urea (SCU), Osmocote® (Scotts, Marysville, OH), Nutricote® (Florikan, Sarasota, FL), and Polyon® (Agrium Advanced Technologies, Sylacauga, AL). Non-coated EEFs include materials of limited water solubility that release plant-available N as they decompose chemically or microbially. Examples include ureaformaldehyde (UF), methylene urea (MU), and isobutylidene diurea (IBDU). In all cases, moisture and temperature play a significant role in determining how quickly N and other nutrients are released.

It is important for a producer to understand how a particular EEF works and to know its designed release rate before applying it in a horticultural situation. Studies evaluating EEF use in vegetable production have emphasized how a crop can suffer if nutrient release is not fast enough. Simonne and Hutchinson (2005) measured N release rates of 18 EEFs tested for utility in growing potato (*Solanum tuberosum*). Only three materials released N at a rate fast enough for potato production; the rest of the materials were too slow. An example of an EEF release curve that is well-matched to plant needs. Citriblen® fertilizer (Scotts) is formulated for use in mature citrus groves, where recommendations suggest that two-thirds of the N should be released within 110 d after applying in the spring.

Timing of EEF application

The goal in EEF application timing is to match the release curve with crop nutrient demand during the growing season (Schwab and Murdock, 2005). For example, newly planted citrus trees should be fertilized at planting, while established trees should receive EEF in late winter or early spring before the first spring vegetative flush and bloom. Turf and ornamental landscapes in southern Florida can receive EEF between April and September, but in northern Florida, it should only be applied between May and July. Vegetables should be fertilized with EEFs preplant, beneath the plastic mulch if applicable (Simonne and Hochmuth, 2007). For greenhouse and nursery crops, EEFs are typically incorporated into the potting media or topdressed after planting.

Placement of EEF

Placement (*e.g.*, surface vs. incorporated) is important to improve NUE and decrease leaching. In an EEF placement study, Broschat (2005) measured lower N, P, and K release rates from surface-applied Osmocote® and Nutricote® fertilizers compared with placing them 10 cm deep in the potting substrate.Million et al. found that surface application of Osmocote® to sweet viburnum (*Viburnum odoratissimum*) did not affect plant growth compared with incorporation, but N and P leaching losses were reduced 16 per cent and 25 per cent, respectively, when the EEF was surface-applied. In a field study, EEFs undergoing evaluation for long-term use on sweet orange (*Citrus sinensis*) trees were applied to the soil surface even if the manufacturer's instructions stated that the material should be incorporated. In all cases, surface application did not detrimentally affect EEF performance, likely because it was applied within the wetted pattern of the microirrigation system that maintained a continuous moist environment on the grove floor.

Temperature and moisture effects on EEF

Temperature and moisture can affect EEF release rate. In most cases, as temperature increases, nutrient release from an EEF increases. However, the temperature-release curve relationship has not been well defined in a quantitative way for many of the EEFs used in Florida. Surface-applied EEFs release nutrients more slowly than incorporated EEFs due to intermittent wetting and drying. For continuous nutrient release, EEF particles need to be continuously moist, but they do not require complete water immersion.

Horticultural performance and N leaching

Numerous Florida studies have measured horticultural performance, and in some cases, leaching, of EEFs. Turfgrass has been the most studied crop in the past two decades.

In a series of field experiments, polymer-coated ammonium sulfate and polymer-coated urea applied to bermudagrass (*Cynodon dactylon*), perennial ryegrass (*Lolium perenne*), and/or st. augustinegrass (*Stenotaphrum secundatum*) produced equal or better turf growth and quality ratings when compared with a standard fertilization programme using WSF. In addition, EEFs reduced N leaching losses by 50 per cent to 75 per cent compared with WSF. However, EEFs have not always performed better than WSF. In a large pot study, Saha et al. (2007) found that even though less N leaching occurred when st. augustinegrass was fertilized with EEF, higher plant quality was observed where WSF was used.

A typical question that a producer considering the use of EEF asks is: "Can I apply a lower fertilizer rate when using an EEF compared with a conventional WSF programme and get the same response?" Environmentally enhanced fertilizer evaluation studies have been structured to answer this question by determining the rate by source interaction in full factorial experiments, but a conglomeration of results suggests there is no definitive answer. Of the 16 experiments summarized, six showed that a lower rate of EEF performed as well or better than a higher rate of WSF, eight showed no rate advantage to EEF, and two showed mixed results depending on the measured response. Of the six "yes" answers, five involved citrus trees and one involved vegetables. Of the eight "no" answers, three involved citrus, three involved turfgrass, and two involved vegetables.

N leaching from field-applied EEFs is assumed to be minimal because, by design, these materials release water-soluble N at a slow rate. Nitrogen not released to the soil remains in an insoluble form or is protected from dissolution by a coating, therefore it cannot leach all at once. Leaching from EEFs has been measured in laboratory simulations with no plants, measured in field lysimeters containing turfgrass, and estimated in a central Florida citrus grove on deep sandy soil.

In all cases, N leaching from water-soluble N fertilizer was greater than leaching from EEFs. Nitrogen leaching from containerized plant production has also been measured. In one experiment, plant growth substrate influenced the relative amount of fertilizer N that leached. Less nitrate leached from EEFs compared with WSF when plants were grown in pine bark–peat–sand media. When the media was sandy field soil, nitrate leaching did not differ between fertilizer types. Phosphorus leaching from EEFs was always significantly lower compared with WSF, regardless of potting substrate type. Plant growth was as good or better with EEFs compared with WSF.

NATURAL ORGANIC SOIL AMENDMENTS AND FERTILIZERS

This grouping includes less processed materials like animal manure and biosolids that may or may not have a guaranteed nutrient analysis, plus more refined materials like fish and seaweed emulsions that are sold as commercial products with a guaranteed analysis. If an organic material is included as part of a nutrient management plan, the mineralization rate must be considered when determining the rate to apply.

Organic matter decomposes relatively quickly in Florida's warm and humid climate. For example, Hanselman et al. (2004) showed that roughly half of the N in biosolids and two-thirds of the N in poultry (layer) manure mineralized to plant-available forms during the first year after application at several Florida locations.

The N mineralization was front-loaded for poultry manure and biosolids (*i.e.*, most of the N became plant-available in the first month after application, followed by a gradual release of the remainder during the subsequent 11 months). A similar release of N from seabird guano, fish powder, feather meal, and blood meal was observed by Hartz and Johnstone (2006). In their study, most of the applied N mineralized within 2 weeks.

FERTIGATION

Nutrient use efficiency can be increased by substituting fertigation for pre-plant WSF application in vegetable production. When bell pepper (*Capsicum annuum*) yield was used an indicator, N use efficiency increased as the proportion of N applied via fertigation increased. Increased yield implies more N in the fruit per unit area, hence less potential leaching. Additional benefits of fertigation for vegetable production are summarized by Locascio (2005).

Fertigation frequency for vegetables can vary from daily application to one fertigation per week. No advantage to daily versus weekly fertigation has been observed with proper irrigation management. Nitrogen fertilizer application is most precise if rates are determined by crop growth and resulting nutrient demand. For example, N rates for tomato and bell pepper begin at 0.5 lb/acre per day during the early part of the season and increase to 2.0 lb/acre per day at peak demand.

Citrus trees do not appear to be sensitive to fertigation application frequency. This characteristic allows wide flexibility when developing a fertigation schedule.

For example, the 1-year growth response of newly planted citrus trees in Gainesville, FL, did not differ when fertigated 30, 10, or 5 times at the same total N rate. Growth observed with fertigation did not differ from that observed with dry granular fertilizer applied five times in 1 year. Fertigation frequency also was not a factor when applied to 6-year-old lysimeter-grown trees in Lake

Alfred, FL. Neither N uptake efficiency nor the relative amount of applied N that leached was significantly different when comparing about 80 fertigations per year with about 12 per year.

Horticultural plant response to fertigation is as good or better than the response observed with well-managed dry soluble fertilization. In either case, irrigation (and sometimes drainage) water management is critical for success. Nitrogen leaching following fertigation can be minimized if the crop is not over-irrigated.

It cannot be overemphasized how important irrigation duration is when fertigating. Although fertigation is sometimes referred to as "spoon feeding," in this case, the "food" is water-soluble plant nutrients that can easily be driven beneath the plant root zone if too much water follows the fertilizer pulse. It is true that fertigation prevents a large mass of nutrients from being leached in a single day (as could occur when heavy rain follows a dry fertilizer application), but leaching can still occur in smaller increments if irrigation management is poor. An interesting result of the study summarized was that even in a lysimeter with carefully controlled irrigation and a confined root system, about half of the N applied via fertigation leached past the root zone.

An example that is contrary to the principle outlined above. Irrigation scheduling in the test citrus grove was optimal, yet more N leached in the fertigation treatment compared with dry soluble fertilizer applications. The authors of this study (Paramasivam et al., 2001) explained that more N leaching occurred with fertigation "purely because of unexpected prolonged irrigation or unexpected high rainfall following certain fertigation events in both years."

FOLIAR FERTILIZATION WITH N AND P

Citrus

The amount of plant nutrients that can be taken up through the leaves of a citrus tree is very small. However, there are special cases where foliar application of N and/or P is justified. It must be recognized that a positive response may be due to additional effects of the materials on tree physiology beyond simple enhancement of tree nutrition.

Forms of urea are available that can be readily absorbed by citrus leaves. Foliar urea sprays applied during the winter have enhanced the number of flowers and yield of 'Valencia' sweet oranges. After cool temperatures or drought stress have occurred, applying 50 to 60 lb/acre of urea can enhance flower bud induction and may increase fruit yield. Maximum penetration of urea into citrus leaves occurs within 12 to 24 h after spray application. Optimum conditions for foliar uptake include air temperature between 77 and 88 °F, high relative humidity, and spray solution with a pH between 7 and 8 to prevent urea breakdown. Under favorable environmental conditions, roughly half of foliar-applied urea penetrates the leaves, while most of the other half is lost through volatilization. The rate of foliar-applied N should be considered as part of the total annual N rate applied to the grove. For example, a foliar spray of 50 lb/acre of urea applies 23 lb/acre of N. If the fertilization plan calls for a total of 180 lb/acre per year of N, only 157 lb/acre should be included in the soil-applied fertilizer programme. In Florida citrus production areas where groundwater nitrate contamination exists or is seen as a potential problem, urea sprays should be evaluated to provide a portion of the tree N requirements, especially during the summer months when leaching potential is the greatest.

Citrus leaves are extremely impervious to phosphate (PO_4^{3-}). Conversely, phosphite (PO_3^{3-}) is more readily absorbed into plant tissue, and once inside the plant, it remains stable. Phosphite does not readily convert to phosphate in the plant, therefore the nutritional value of absorbed PO_3^{3-} is uncertain. However, phosphite is officially recognized by FDACS as a source of P for crops. In Florida, a prebloom foliar application of 2.6 qt/acre of 28 per cent P_2O_5 as potassium phosphate (K_3PO_3) to 'Valencia' sweet oranges significantly increased flower number, fruit yield, and total soluble solids yield compared with an untreated control. These results suggest that the effect of phosphite was not due to the molecule's fungicidal attributes, but to other growth-stimulating properties.

Other horticultural crops

Foliar applications of N and P are not recommended for vegetable production because leaves cannot absorb sufficient quantities to correct a deficiency, and leaf burn is likely if this is attempted. This principle also applies to ornamental plants and turfgrass.

EXAMPLE COST COMPARISONS OF EEF VERSUS WSF PROGRAMMES

Citrus

Although coated EEFs performed very well in a 6-year trial comparing them with a standard WSF programme (1991–96), they would not have been economically feasible for commercial production. The EEF materials evaluated would have cost three to four times as much to use as WSF, even when the lower application cost was factored in. Extra yield obtained by using EEF did not nearly make up for the higher fertilizer cost.

Vegetables

A potato production study determined that the cost of a water-soluble N fertilization programme in most years would fall between \$72/ha and \$158/ha. Estimated EEF programme cost would be about \$20/ha to \$195/ha more than the most expensive soluble N cost. This extra cost could be offset by reduced application cost and/or providing cost-share for the use of EEF.

HOW IRRIGATION AFFECTS FERTILIZER USE EFFICIENCY

Irrigation effects on fertilizer NUE for vegetable production were summarized by Hochmuth (2003) and Locascio (2005), and for citrus production by Obreza and Morgan (2008). Fertilizer use efficiency of container-grown ornamental production was improved by changing the irrigation and/or fertilization methods. One change involved the conversion of overhead irrigation to microirrigation or capillary mat/wick irrigation. The other change involved the use of EEFs [Osmocote®, Nutricote®, Multicote® (Haifa Chemicals, Haifa, Israel), and Polyon®] with overhead irrigation. During a 4-year period, the growth index of peace lily (*Spathiphyllum* spp.) was about the same regardless of EEF source. However, the nitrate-N concentration in the groundwater 4 ft below the ground surface of the shadehouse decreased as a result of EEF use. In another study involving containerized production of Osmocote®-fertilized sweet viburnum, excessive irrigation (double the required rate) decreased plant growth, increased run-off volume from the nursery, increased N loss in run-off by 21 per cent to 34 per cent, and increased P loss in run-off by 28 per cent to 38 per cent.

FUTURE RESEARCH NEEDS

Despite numerous improvements in fertilizer use efficiency that have occurred during the past several decades, further advancement is possible if research is undertaken on the following aspects of fertilizer use: 1) develop a short-term procedure to verify the nutrient release period claimed on EEF labels, 2) evaluate nutrient leaching characteristics of EEF materials applied in the field and greenhouse, 3) conduct an economic study of EEF use that determines the value of the environmental benefit, 4) improve irrigation scheduling techniques, and 5) develop new technology for precision nutrient application.

7

Increase Water Use in Crop Production

INTRODUCTION

The supply of some freshwater is an absolute essential for all forms of agriculture, although the amount of water required varies greatly between different agricultural types and climatic regions. Given the scale of agricultural activity, this means that agricultural activity dominates the use of freshwater and accounts for some 70 per cent of withdrawals from water resources globally. Such large abstraction of water has led to large reductions in river flow, indeed, to long and repeated periods of zero flow for several major river systems, with consequent huge changes in estuarine and coastal ecosystems. In addition, abstraction has overexploited many major regional aquifers. As most of the water withdrawn by agriculture is lost in evaporation and transpiration, in marked contrast to domestic and industrial withdrawals, agriculture accounts for 80–90 per cent of freshwater used by humans. Clearly, making agriculture sustainable requires a major reduction in water use in many regions.

Shortage of water leads to drought with obvious agricultural and societal impacts. Periods of repeated droughts have a particularly severe effect on societies, and there are many historical examples. For example, the fall of the Roman Empire has been attributed to a drying climate, both in Europe and in the Central Asian steppes, which may have stimulated the movement of the barbarian tribes.

The decline of the Maya civilization has been attributed to droughts of increasing frequency and severity between AD 750 and 1050. More recently, the difficulties of British settlement in North America in the beginning of the seventeenth century may have resulted from coincidental extreme droughts. In current times, we are all too familiar with droughts and consequent famine in Africa, for example. Furthermore, there is widespread agreement that increasing anthropogenic climate change will exacerbate the present shortages of water, and is likely to increase drought.

Farmers in many countries are now faced with legislative restrictions on use of water, which are being imposed to try to secure safe and adequate water supplies for domestic water users. Under some circumstances, restrictions on use can be substantial. For example, in China, government has set a target of a reduction of 20 per cent in water use in agriculture by the year 2020. If food security for the region is not to be threatened, this must be achieved without a loss in production.

Overuse of water in some regions of China is having serious environmental consequences. For example, the Shiyang River Basin in arid, northwest China is the largest inland river basin in the region with the largest human population density and the highest exploitation of water resources. Over the last 50 or more years, as a result of an explosion of agriculture in the region, severe water shortage has developed which not only constrains social and economic developments but is also leading to some of the worst ecological and environmental deterioration in China. Kang *et al*. have highlighted the changes in the water systems in the Minqin oasis at the end of the river basin caused by both the changing climate and intensive farming activity.

With an increasing population (159 per cent rise in 50 years), the amount of cultivated land in the basin has been expanded by more than 50 per cent. Many reservoirs have been built by damming rivers and large-scale irrigation has been introduced.

Water use due to human activities has exceeded the capacity of the water resources in the basin, which has led to a dramatic and rapid fall in the water table in the Minqin oasis (by as much as 14m). The oasis is shrinking in area, natural vegetation that relies on underground water is disappearing and desertification is accelerating, leading to more dust storms.

'Super' dust storms occurred only five times in the 1950s, but in 2001 they occurred six times in the first half of the year. Now the Minqin area is one of the origins of major dust storms for China and the rest of the world. With the rapid deterioration of its water resources, increasing droughts and dust storms and with large areas of abandoned land, more and more farmers are eventually migrating out as 'ecological refugees'. So, an ancient oasis that can be traced back 2000 years is disappearing, with local and global effects on the society and climate.

Thus how to reduce agricultural water use and make water resources more sustainable is an increasingly urgent question. It is a question that requires combined agronomic, physiological, biotechnological/genetic and engineering approaches which may be collectively described as 'water saving agriculture'. We need to use appropriate agronomy to grow crops best suited to the environment, with least use of water, whether in irrigated or rainfed production. We need to develop crops that require less water to produce sufficient yield, through understanding the physiological mechanisms that determine growth and water loss, and plant response to reduced water availability. We need to consider fully the returns on irrigation, as part of making real improvements in sustainable water management. Ideally, as many commentators have pointed out, we need to reduce the use of irrigation in hot, dry environments, as it is in these environments that irrigation is most wasteful, in that it produces least yield per unit of water, owing to the high evaporation rates. Similarly, some techniques are less efficient than others.

For example, in a recent survey in England, 72 per cent of farmers irrigating field crops were using rain guns, despite their inefficiency in delivering water to the crops. Ultimately, the goal must be to reduce the amount of water used per unit yield achieved. As Kofi Annan, UN Secretary General, declared in 2000 'we need a Blue Revolution in agriculture that focuses on increasing productivity per unit of water—'more crop per drop'. Perhaps it would be most beneficial if the maximum nutrition per unit of water was achieved, but economics dictate that crops grown are those that produce the maximum economic yield, so that the exact definition of sustainability becomes important. The amount of yield produced per unit of water used is referred to as 'water use efficiency', or 'water productivity', and a discussion of how this might be improved is the focus of this chapter.

WATER USE IN AGRICULTURE

Some figures at different scales illustrate the amount of water used in agriculture. Globally, some $2.7 \times 10^3 km^3$ of water were used in agriculture in 2000. While significant amounts of water are used in animal production and food processing, water availability limits animal production mostly through the production of fodder crops and forage, and it is therefore crop production that we consider in this paper. Figures that are widely quoted are that production of 1kg of wheat requires $1m^3$ of water and 1kg of rice requires at least $1.2m^3$ of water.

However, these are low requirement figures compared to some, such as the $0.7–2m^3$ required for grain yield in temperate zones and $3–5m^3$ in arid regions given by Gregory (2004). Most of the water used is for the irrigation of some 276Mha of crops. The area irrigated is still increasing globally, although the rate of increase in area irrigated appears to be reducing, averaging 1.1Mhayr"^{1} between 1999 and 2003, compared with 4.2Mha yr"^{1} between 1990 and 1994 (data from FAOSTAT 2006). In the 15 countries of the EU in 2003, an area equivalent to 15.5 per cent of the arable and permanent crop area was irrigated and irrigation comprised over half of the total water consumption. Even within the humid, temperate climate of England, 147kha of outdoor crops were irrigated in 2001 (approx. 3 per cent of the cropped area) using 131×10^6 m^3 of water.

The key point about this irrigation demand is that although irrigated areas accounted for approximately 18 per cent of the world's cropped land in 2003, they produced approximately 40–45 per cent of the global food. Therefore, reduction in irrigated areas or productivity could have very serious impacts on global food supply. Conversely, any reduction in irrigation water used would have major benefits to freshwater resources. However, the other 82 per cent of the global cropland which is rainfed and the 3.4bn ha of rainfed permanent pasture is just as important as it is obviously a major part of the local catchment and regional hydrological balances, and a major contributor to local and regional food security. Even in the Mediterranean region, with large irrigated areas, 60 per cent of the cereal production comes from rainfed cropping.

Clearly, at a regional scale, the major problems in water supply are in regions with low rainfall, yet high evaporative demand, and those with expanding populations, such as North Africa, Southern Africa and the Near East, where water availability is less than 2000m^3 per person per year. Population growth, urban expansion and economic development are increasing other demands for water. Therefore, there is increased competition for water between agriculture and other users, and in many regions reductions in water quality as well as quantity. Salinization and other contamination of surface and groundwater are major problems, although outside the scope of this paper. Agricultural water use is also a major international political issue, for example, in regions where irrigation depends on the river inflow from neighbouring countries, such as in Egypt, Iraq or Syria.

In addition, most plausible predictions for the climate over the next 50 years indicate increases of annual mean temperature of 1–2.5°C, but possibly considerably higher. This will probably increase evapotranspiration rates, particularly in warmer regions, and T is transpiration rate. Precipitation amounts are also changing, and in many temperate environments there are already changes to the seasonal distribution, with further predictions of milder, wetter winters and warmer, drier summers for much of Europe, for example. Other serious changes predicted are less snowfall for key melt-water rivers that provide water for major crop growing regions, such as in India and Pakistan, and for more storms and localized flooding.

In addition, there are predictions that sea-level increase and greater abstraction of water will increase salinization of coastal aquifers. All of these changes add substantially to pressures on water resources and increase the urgency of reducing water use in crop production, thus improving yields and yield stability at the same time.

NEW TECHNIQUES FOR ASSESSING PLANT TRANSPIRATION EFFICIENCY

In the 1980s, Farquhar *et al.* (1982) developed a new method for assessing *TE* using the ratio of the abundance of the natural isotopes of C, ^{13}C and ^{12}C. During diffusion and biochemical fixation of CO_2, the heavier ^{13}C is discriminated against, so that inside leaves and in leaf tissue the ratio $^{13}C/^{12}C$ is reduced compared with the normal abundance in the atmosphere. The ratio (or ä^{13}C, the discrimination against ^{13}C) depends on the balance between diffusion into the leaf and demand by *A*, so ä^{13}C gives a measure of *TE*, either at the moment (for gas measurements) or integrated over the period of fixation of leaf organic material. It is therefore related to the ratio of A/g_s and to *ATR*.

Drier conditions usually result in less discrimination (less negative ä^{13}C values). The approach has now been used to investigate *TE* in many crop and tree species, ranging from cereals, through cotton and soya bean, to grapes, coffee, eucalypts and pines. The discrimination in leaf material of 15 coconut genotypes was found to be very closely related to the A/g_s measured by gas exchange. In such C3 plants where there is a strong relationship between C_i and ä^{13}C, the technique has led directly to the selection of improved crop varieties, most notably in wheat. However, it is not suitable for screening C4 plants because ä^{13}C also depends on the 'leakiness' of the bundle sheath for CO_2.

Furthermore, the ä^{13}C method cannot distinguish between changes of net assimilation rate or of stomatal conductance. More recently, the isotopic ratio of ^{18}O to ^{16}O has been used to indicate the transpiration conditions during C fixation. In a similar way to the C isotope discrimination, the ratio of the oxygen isotopes in leaf water or in leaf organic matter depends on diffusion and transpiration rates. The heavier isotope becomes enriched at the sites of evaporation because ^{16}O-containing water is preferentially evaporated, resulting in a positive discrimination, but the relationship with conductance is altered by leaf temperature and air humidity which affects the gradient in vapour pressure. Therefore, when ä^{18}O and ä^{13}C data are combined, it can provide information on the relative contribution of stomatal conductance or CO_2 assimilation rate in determining *TE*.

Much of the routine use of the above isotope discrimination approaches has been based on leaf or seed material. This material is dominated by the cellulose fraction and so integrates the conditions during and after organ growth, and can be rather insensitive to changes in photosynthetic rate caused, for example, by drought stress after expansion. It is possible to refine the technique to look at the dynamics of photosynthetic fixation as stresses occur, by examining just the soluble and starch carbohydrate components, but this is not yet routine.

Clearly, techniques are required for rapidly assessing both the photosynthetic rate and stomatal conductance of leaves for the screening of genotypes and their response to drought or other environmental stresses. Chlorophyll fluorescence emissions from intact plants are easily and rapidly measured, and have been shown to be directly related to photosynthetic activity. The fluorescence parameter F_q2/F_m2, which estimates the quantum yield of photosystem II photochemistry, has been shown to be directly related to the rates of linear electron transport and CO_2assimilation in leaves. Consequently, this parameter potentially provides an effective tool for rapidly screening for differences in photosynthesis and thus A, or for differences in the response to stresses like drought.

However, it is well established that photosynthetic performance across the area of individual leaves can be extremely heterogeneous, especially for stressed plants. Any attempts to screen for effects on leaf photosynthesis using fluorescence should ideally sample the whole of the leaf. Conventional commercial fluorimeters generally only measure fluorescence from a small area of leaf and are unsuitable for such screening. Imaging instruments using charge coupled device (CCD) cameras overcome this problem and allow an image of fluorescence parameters to be produced from whole leaves. An example of the heterogeneity in leaf photosynthetic function caused by drought.

Another imaging technique, IR thermography (IRT), has recently been used to study stomatal responses to drought stress. IRT can be used in a simple way to look for differences in leaf temperature, and thereby infer differences in transpiration and stomatal behaviour. It has been used this way to screen for and identify mutants with altered stomatal function. *Arabidopsis thaliana* (*Arabidopsis*) plants in a tray, with the thermal image showing that the leaves of the OST plant (circled) were 1–1.5°C cooler than those of surrounding wild-type plants.

In addition, IRT can be used to estimate g_s for whole leaves and canopies if suitable reference surfaces are used. An index I_g that is proportional to g_s can be readily determined from thermal images of leaves and is a suitable parameter for rapidly screening for genotypic variation in g_s. This approach with the*Arabidopsis* example. Horie *et al.* (2006) have recently shown the strong potential of IRT for selecting for rice genotypes with higher conductance associated with higher yields.

By coupling fluorescence and thermal imaging techniques it is now possible to rapidly evaluate both the photosynthetic performance and stomatal conductance of leaves. This has been done at the small scale as a tool to examine stomatal control of photosynthesis in leaves. Clearly, as the ratio of the two parameters $(F_q2/F_m2)/I_g$ is directly proportional to A/g_s, combining such images offers exciting possibilities to screen plant material for photosynthesis, water loss and ATR non-destructively and repeatedly.

DROUGHT RESISTANCE AND WATER PRODUCTIVITY

One approach to understand plant water productivity is to study plants from dry environments which have mechanisms that confer drought resistance, traditionally divided into dehydration *avoidance* and dehydration*tolerance* mechanisms. However, it has been argued that such comparative studies are rarely useful as drought resistance is usually linked to low productivity, and is thus of limited use in agricultural production. In addition, severe drought is usually prolonged, so survival will not usually lead to subsequent productivity, as the water is simply not there. Also, Passioura (2002) has stated that 'severe water deficits are rare in viable agriculture, and asking how crops respond to or survive them is unlikely to have much general impact'.

This is true for commercial agriculture, but is perhaps not true of much of 'subsistence' farming. In these situations, it is possible that physiological mechanisms that allow maintenance of a high plant water status such as osmotic adjustment, reduced water loss through cuticle, avoidance of xylem cavitation and altered root-to-shoot ratio could be important in providing some yield in 'resource-poor' cropping systems.

Several recent reviews have pointed out that the main improvements in drought resistance of crops have come from increasing dehydration avoidance, particularly by increasing *W*, through changes such as earlier development, smaller leaves, and deeper roots, and not by improving dehydration tolerance.

One important example of improved dehydration tolerance is in improving the response of arable crops to terminal drought by improving the mobilization of stem carbohydrate reserves for grain filling. However, in many regions where drought stress occurs, rainfall is variable, and whether increased drought resistance will inevitably result in plants that will not be able to take advantage of any higher rainfall needs to be further investigated. Blum (2005) points out that there is usually a negative relation between yield potential and yield drought resistance, and argues that genotypic variations in *BWR* are normally due to variations in water use, and improvements are usually achieved by traits and environmental responses that reduce the yield potential. Blum therefore cautions that selection for *BWR* directly in order to improve drought resistance is not necessarily the best way forward.

Furthermore, Parry *et al.* (2005) have pointed out that most biotechnological approaches using transgenics to improve performance under drought have targeted survival and reductions in water use, not improved photosynthesis under a given supply of water. Passioura (2004) has noted that improving water productivity is now seen as the way forward, rather than the search for drought resistance.

CROP MANAGEMENT STRATEGIES

There are a number of strategies for crop management that can improve water use efficiency. These have been reviewed in several recent papers. In brief, they can be grouped into the following.

a. Capturing more water (*W*) for crop transpiration through water harvesting, reducing soil evaporation, improved weed control and by deeper root growth.

b. Improving *BWR* by exchanging transpired water for CO_2 more effectively and converting into biomass.
c. Convert more biomass into harvestable yield (*HI*).

Water harvesting approaches may be appropriate in some dryland regions, by harvesting water on larger areas and channelling it onto crops. This is, in effect, a form of irrigation that has been practised traditionally in some regions. However, it is questionable that this is sustainable when viewed on the wider landscape scale, as it reduces groundwater recharge and possibly seasonal river flows that may be important to catchments and natural ecosystems.

Soil evaporation can be substantially reduced through minimum tillage and mulching using materials or crop residues. Most importantly, there is considerable recent work showing that growing paddy rice without standing water and with soil mulches can provide major improvements in water use efficiency, with small or no yield penalties. Traditional paddy rice cultivation has used inundation for some 5000 years, but this results in 2–5 times more water use than other cereals.

Given that paddy rice cultivation now covers a huge area and is of overwhelming importance in global food supply, these are potentially major developments, although the full agronomic, environmental and socio-economic consequences of such changes are complex. Soil evaporation can also be reduced by achieving a rapid ground cover through early crop vigour and fertilizers have been shown to help. However, there is a water cost to increased ground cover: the larger crop area will transpire more and thus needs to be coupled to selection of suitable genotypes to avoid later season water shortages.

Improving *BWR* can potentially be done by improvement of the genotype. In addition, reducing yield losses from weeds, pests and diseases through mechanical and agrochemical methods can achieve improved *BWR*, although such practices raise other questions about sustainability. However, altered management interacts with physiological changes introduced in new genotypes.

For example, much has been achieved by selection of crops or cultivars that grow earlier in climates such as Mediterranean regions where potential evaporation increases during the season, and Ä*e* is lower earlier on. This close matching of phenological development with rainfall patterns and environment is very important, but needs to be coupled with management that ensures early and rapid sowing. Adoption of minimum tillage practices should also help with early and rapid sowing. As a result of such better matching of crops and environment, coupled with the introduction of cultivars that respond to increased inputs, improved disease control and other agronomic changes, yields have doubled over the last 50 years in rainfed agriculture in Australia. Consequently, water productivity has doubled and both of these improvements have occurred even though water available for agriculture has declined over this period.

The *HI* of many of the major crops has been substantially improved genetically, and it is sometimes argued that the *HI* of cereals is reaching an upper limit in some systems. New improvement targets are for improved assimilate production, storage and remobilization at the optimum time for seed filling, as in the stay-green sorghum lines with *delayed* senescence. In contrast, management techniques such as deficit irrigation can be used to *promote* senescence, to promote remobilization of assimilate from storage in the stem to increase grain yield in both wheat and rice. In these species, stored non-structural carbohydrates in the stem and leaf sheaths may contribute 10–40 per cent of the final grain yield, and this may be unavailable for grain development unless senescence is promoted. The problem may be particularly acute if there is an excess of nitrogen in the soil after anthesis.

There is substantial evidence for all these measures contributing to improved crop yield in Mediterranean-climate areas, and there are good examples in other climates. Recently,Zhang *et al.* (2005) have reported that there have been 50 per cent yield and water use efficiency increases in the North China Plain in winter wheat and maize over the last 20 years associated with mulching and improved irrigation scheduling. In the Central Plains of the USA, no-tillage practices have made it possible to intensify cropping from the traditional wheat–fallow system and produce a 30 per cent increase in water use efficiency, as well as increase soil C.

DEFICIT IRRIGATION AS A MEANS OF MANAGING CROP WATER LOSS AND INCREASING HI

John Boyer showed in 1970 that in many cases, shoot growth is very sensitive to mild soil moisture deficits, and more sensitive than photosynthetic rate. Owing to this shoot growth sensitivity, it is common in many production systems to supply excess irrigation water to ensure rapid and extensive canopy development and thereby maximize radiation interception and biomass accumulation. In many situations, irrigation systems are badly designed or maintained so that much of the water goes nowhere near the crop. The spatial and temporal distribution of water is highly variable and growers will often leave irrigation systems running to ensure that the driest plants receive enough water to avoid the development of damaging water deficits. Thus, the wettest plants receive perhaps 3–4 times more water than they need, which itself can have damaging effects on crop growth and development.

There is now much interest in using remote sensing techniques such as thermal imaging to assess the level of plant stress and to use such measurements to control irrigation systems. This combination should result in water being applied to the plants that need it, in just the required amounts. Such technology which uses the plant as an indicator of water requirementhas the potential to greatly reduce the use of water in commercial, highly mechanized agriculture and very significantly increase *BWR*. One management option with great promise is *deficit irrigation* (the application of only a predetermined percentage of calculated potential plant water use), and in particular, regulated deficit irrigation (RDI). As less water is supplied, mild soil drying occurs, but stomatal aperture is reduced more than assimilation rate (owing to the non-linearity of the ratio A/g_s) so that *ATR* is increased, and possibly *BWR*. The mild soil drying resulting from the use of deficit irrigation will almost inevitably restrict shoot growth and leaf development.

This response to drying may have several benefits for growers. Firstly, many fruit crops can show excessive vegetative vigour, which can lead to high rates of water loss and development of damaging plant water deficits. Secondly, leaf growth can also compete with reproductive development for assimilates.

Thirdly, excessive canopy development can cause shade, predispose plants to fungal diseases and adversely affect fruit development and fruit quality. While vigorous crops are often pruned, this can provide infection sites for diseases and can be expensive if labour costs are high. Therefore, deficit irrigation can be beneficial in fruit crops. In cereals, it may also result in beneficial redistribution of assimilates from storage to developing grains to increase HI and grain yield. However, deficit irrigation can also reduce fruit yields with the risk of catastrophic crop failure, so that to operate RDI effectively, the grower must monitor plant and soil moisture status and deliver precisely regulated and directed quantities of water on demand.

PARTIAL ROOT ZONE DRYING TO MANIPULATE THE PLANT'S SIGNALLING CAPACITY

Plant growth in drying soil is commonly limited by a combination of chemical and hydraulic influences. Frequently, reductions in water availability result in reduced shoot turgor which can reduce shoot growth and development. Even when turgor of growing shoot cells is sustained, growth can be limited by chemical 'signals' generated as a result of interactions between the root and the drying soil and transmitted to the shoot via the transpiration stream. Gowing *et al*. (1990) showed that production and expansion of leaves of apple trees could be restricted by watering only half the plant root system (purtial root drying; PRD) and that this limitation occurred without any detectable influence of the treatment on shoot water status. Excision of the roots in the drying soil promoted shoot growth so that the growth limitation under PRD was attributable to delivery to the shoots of chemical inhibitors in the transpiration stream, generated as a result of soil drying.

In recent years, there has been much interest in modifying the long-distance signalling through deficit irrigation and thereby modifying plant growth to the advantage of the producer and saving water. This possibility was first mooted by Loveys (1991), who suggested reduced water application to grapevines to restrict vegetative shoot development. It was proposed that irrigation could be regulated so as to control stomatal conductance, reduce plant water loss but sustain shoot water status to ensure 'normal' fruit development. The fruit turgor of tomatoes can be sustained under a particular form of deficit irrigation, and Dry and Loveys (1999) have described results from experiments where grapevine vigour was significantly reduced and less water used without any yield penalty. Fruit production per unit of water used was therefore greatly increased compared to the industry standard. In addition, the quality of fruit was improved significantly.

In PRD, the two halves of the plant's root system are watered alternately. Roots in wet soil supply most of the water, while roots in contact with drying soil generate signals that move to the shoot to restrict shoot growth and functioning. Irrigation must be switched regularly from one side of the root to the other to keep roots in dry soil alive and fully functional and sustain the supply of root signals. Maintenance of turgor and total water potential in the shoot can be important for maintenance of fruit growth and development although it is not entirely clear why fruit growth does not respond as sensitively as leaf growth to the increased flux of chemical signals. Increases in grape quality combined with reduced water use have now been reported as a result of the use of PRD on a commercial field scale.

The application of the technique is not restricted to grapevines or other tree crops. In an extended series of studies in Australia, the UK, China and in countries around the Mediterranean, application of PRD to a range of other crops with different irrigation methods has resulted in substantial saving of irrigation water coupled with the maintenance of an economic yield. In these experiments, *BWR* has often at least doubled and in some cases increased much more. Despite some difficulties in operating PRD irrigation techniques in certain soils and some climatic conditions, it is clear that substantial water savings and increases in crop quality can result. Large-scale development of the technique continues.

CONCLUDING COMMENTS

Water is a key driver of sustainability. Crop production in many parts of the world is seriously limited by lack of water, and water supplies for agriculture are also dwindling. Ultimately, this threatens the sustainability of agriculture in these regions, and furthermore agricultural water use has major impacts on the rest of the ecosystem. In areas where irrigation is used, there is an urgent need to increase the efficiency of supply and make much better use of the water applied. In other areas, rainfall is the only source of water for crop growth, thus in those areas any improvements in food yield per unit land area through crop management, fertilization, soil improvement, pest and disease control will usually improve water productivity. Many of the innovative local and regional schemes in developing countries where techniques have been adopted to improve agricultural system sustainability have improved yield and yield stability, and thus often improved water productivity.

While major efforts are being made to improve yield per unit of water supplied, we clearly need more progress. As this review has demonstrated, many interlinked processes and factors underlie the apparent simplicity of the basic requirement for water in crop production, as encapsulated in equation. This inevitably means that to make progress requires understanding and progress in all of agronomy, hydrology, agricultural engineering, crop and plant physiology and molecular genetics.

In addition, where changes are being made to fundamental aspects of a cropping system, such as changing the inundation duration of rice, the socio-economic aspects are as important. Encouragingly, there is currently exciting progress in improving crop production and water productivity through a combination of these disciplines as the examples described here show. Many of the more technologically demanding approaches are unlikely to be appropriate or available to small farmers. However, the development and release of new varieties with characteristics that will improve water productivity through combined physiological, biotechnological and agronomic research will help with the 'Blue Revolution' that is being called for, which could benefit all crop production.

COMPUTATION OF WATER REQUIREMENTS FOR CROP PRODUCTION

To estimate the effect of the water management options on securing global crop production, we followed the approach of Rockströmet al, assuming that as a global average 1300 m3 cap-1 yr-1 of blue and green water are required under present conditions to produce 3000 kcal cap-1 day-1 for a healthy diet (20 per cent meat share).

Actually, however, this water requirement differs among regions. To roughly account for these differences, we distributed the global average among countries according to their water productivity (*i.e.* the water consumed per unit of NPP), such that regions with low water productivity got values higher than 1300 m3 cap-1 yr-1(*e.g.* around 2000 m3 cap-1 yr-1 in many countries of sub-Saharan Africa), and vice versa.

Countries in which the water availability per capita falls below the thus computed value are regarded as water-stressed. Analogously, the water stress threshold changes under each of the scenarios computed here, following the individual changes in NPP and E and, thus, water productivity.

MODEL SETUP

LPJmL was driven by an updated version of the CRU TS2.1 climate dataset

(0.5° spatial resolution) of monthly air temperatures, precipitation, and cloud cover for the period 1901-2000, preceded by a 1000 yr spinup during which the climate of 1901-1930 was repeated.

CO_2 concentrations were as in Rost *et al.* For the simulations under future climate, starting from the same equilibrium, the model was driven by the output from three different climate models under the A2 emissions trajectory for the period 1901-2070.

Sub-monthly weather variability was derived by statistical procedures. All results were computed at daily time steps for each CFT separately, but are presented as annual averages over all CFTs for 1971-2000 and 2041–2070, respectively.

PRESENT WATER LIMITATION OF CROP PRODUCTION

We estimated a maximum achievable global crop NPP* of 23.5 Gt in the absence of water limitation (OPT simulation). Only about half of this theoretical potential would be realised without any irrigation.

This reflects the fact that NPP is strongly water-limited in many regions of the world, and suggests a high potential for increasing NPP through proper management of blue and green water in water-scarce regions.

In contrast, in tropical, marine west coast, humid subtropical, and subarctic climate zones NPP (INO) is already close to NPP*, which implies low potentials for increasing NPP.

POTENTIAL OF WATER MANAGEMENT STRATEGIES UNDER PRESENT CLIMATE

The potential to increase global crop NPP by reducing soil evaporation is estimated to be 0.31 Gt (2.3 per cent) under low (VS10) and 3.3 Gt (24.7 per cent) under intense (VS85) vapour shift management.

The VS25 simulation suggests a high potential (> 20 per cent) mainly in semiarid regions such as the midwestern United States, parts of South America, the Sahel, southern Africa, Central Asia, and south-eastern Australia Even on irrigated areas, vapour shift management can reduce water requirements by 5 per cent (147 km3 yr-1) under low (VS10) and by 83 per cent (1499 km3 yr-1) under intense management (VS85) (data not shown).

Together with improvements in irrigation efficiency not studied here-which may, together with expansions of irrigated areas, amount to about 800 km3 yr-1 globally -the thus saved blue water could be used for purposes other than agriculture, or for expanding irrigated areas to currently rainfed cropland. Run-off harvesting was simulated to increase global NPP by 0.57 Gt (4.2 per cent) under low (RH10) and by 4.1 Gt (30.7 per cent) under high storage (RH85). In the RH25 simulation tropical wet and dry regions like Central America, parts of Brazil, eastern and southern Africa in particular show large potentials These magnitudes are comparable to those in Wisser *et al*, who found that overall cereal production in low-yield regions could be increased by about 15 per cent for a medium variant of small reservoir construction, with highest potentials in Africa and Asia.

Regional differences in the potential to increase NPP between the VS and RH scenario can be explained primarily by climatic differences. Whereas vapour shift management seems to be more efficient in dry regions with low R, run-off harvesting was found to be more effective if substantial amounts of R can be stored outside of the growing period. In addition, crop management practices depend strongly on soil texture, with improvements in crop production being most promising on soils with high water holding capacity.

The combination of both management strategies (VS + RH) results in an increase in NPP by 0.92 Gt (6.9 per cent) in VS10 + RH10 and by 7.05 Gt (52.8 per cent) in VS85 + RH85. Pronounced increases in NPP can be achieved mainly in regions where present NPP reaches less than 10 per cent of NPP*, which agrees well with recent findings by Faures and Santini for sub-Saharan Africa.

The aim of this study is to demonstrate the range of NPP increases given different levels of water management intensity without deeming a particular scenario to be most realistic. Nonetheless, we think that a global implementation of the 85 per cent scenario is clearly unrealistic (though physically possible), and even the 50 per cent scenario is very ambitious, while the 25 per cent scenarios seem to be a possible option: Falkenmark and Rockström suggested an improvement of green water productivity by 1530 km3 yr-1 through a combination of various techniques, which is comparable to our estimates of a water saving of about 1650 km3 yr-1 and an associated NPP increase of 2.5 Gt in the V 25 + RH25 scenario.

The actual potentials and mixtures of VS and RH methods will certainly differ between regions and require investigation in dedicated regional studies. Note that we did not account for evaporation losses of harvested water nor for the possibility that water harvesting reduces blue water availability in downstream locations, which may lead to lower overall water savings within river basins and globally, especially in the scenarios of high-level management

FUTURE POTENTIAL OF WATER MANAGEMENT UNDER CLIMATE CHANGE ONLY

The effect of climate change only (CC) is a decrease in NPP*, and in NPP under all water management simulations, compared to the present situation. This is due primarily to regional precipitation declines and generally higher temperatures that lead to higher crop water limitation and more direct crop damage.

Similar results were found, who estimated a climate change-driven decrease in global crop production of 10 per cent by 2080. Under climate change only, irrigation on areas currently equipped for irrigation would increase the otherwise rather low crop production by 19.5 per cent from 10.2 to 12.2 Gt, *i.e.* future irrigation will continue to play an important role on present land equipped for irrigation.

The potential to increase NPP by the other water management strategies ranges from 0.26 Gt (2 per cent) in VS10 to 3.9 Gt (32 per cent) in RH85, and from 0.82 Gt (7 per cent) in VS10 + RH10 to 6.5 Gt (54 per cent) in VS85 + RS85. The absolute rise in crop production that can be attained was found to be slightly lower under future climate than at present, due to globally lower amounts of water stored in the soil and/or harvestable (data not shown). The relative potential to upgrade NPP by water management, however, is of a similar magnitude under future climate change as at present.

Note that either a vapour shift of about 50 per cent or a water harvesting level of 25 per cent would have to be achieved if the adverse consequences of global climate change were to be counterbalanced, *i.e.* if the current crop production levels were to be held. However, crop production will have to be increased significantly in the future, with the actual amount of this increase depending on population development.

FUTURE POTENTIAL OF WATER MANAGEMENT UNDER CLIMATE AND CO_2 CHANGE

Other than under CC, the global joint effect of climate and CO_2 change (CC + CO_2) is an increase in NPP* and NPP, respectively. The isolated effect of rising CO_2 concentration is an increase in global crop NPP by 28 per cent by the 2050s, which is somewhat higher than suggested by forest FACE experiments (about 23 per cent at a CO_2 level of about 500 ppm;. Irrigation was simulated to increase NPP by 16 per cent to 17 Gt. The potential to increase NPP by the diverse water management options ranges from 0.26 Gt (1.5 per cent) in VS10 to 3.7 Gt (22 per cent) in RH85, and from 0.78 Gt (5 per cent) in VS10 + RH10 to 6.3 Gt (37 per cent) in VS85 + RH85.

It is noteworthy that the CO2 effect more than offsets the global NPP decrease induced by climate change, as was also shown elsewhere. At any rate, the beneficial CO2 effect should be interpreted as a maximum effect, the realization of which represents a major challenge due to the complex interactions between photosynthesis and yields, and limitations to crop growth through low nutrient availability, soil degradation, pests, weeds, and diseases.

These co-limitations, and especially N and P limitations, are not explicitly considered in our model, but, as concludes, current observational evidence suggests that the projected strong impacts of increasing CO2 on NPP are realistic if other limitations are absent. Please also note that we assumed the present management intensity to remain constant in the future, meaning that absolute NPP may be higher (or lower) in the non- CO2 case. In this context we recognize that the computed ET:E ratio-which is influenced by management and which, in turn, influences the outcome of the VS simulations in particular-depends on the daily phenological status of the CFTs. Preliminary studies based on more recent land use data and a refined parameterization of agricultural management show, however, that the present results are still valid.

WATER USE AND WATER-USE EFFICIENCY

It should be realised that various expressions for water-use efficiency are used, and that the physiological and agronomic definitions may differ. Time-averaged water use and water-use efficiency (seasonal or periodic whole-plant transpiration rate (WPTR), and water-use efficiency (WUE)) are often confounded with instantaneous transpiration rate and transpiration efficiency (ITE). The latter expression is best described as the ratio between the instantaneous CO2 fixation (actual net photosynthesis) and water loss by transpiration (both measured with a leaf or plant chamber). Water-use efficiency is the ratio of the net gain in dry matter over a given period, divided by the water loss (from the vegetation alone or from soil and vegetation together) over the same period. Mostly the ITE, normally only measured during the light period, is higher than WUE, both because of variations in ITE over day and season and because of carbon losses by respiration in the dark. WUE seems to be the more relevant parameter to the question of impact on agricultural production.

INTERACTIONS WITH EXPERIMENTAL CONDITIONS

Most available data on water use and water-use efficiency have been collected from single plants grown in pots or hydroponically, and using porometer measurements supplemented with periodic weight determinations. Values for water-use efficiency and total water use in (semi-) field conditions over the full season, or an extended period, are scarce. Moreover, a complication for the correct estimation of the effect of elevated [CO_2] and temperature changes on transpiration is that almost all experiments in this field have been performed in environmentally controlled and generally well mixed and ventilated experimental set-ups (enclosures or Open-Top Chambers (OTCs)), where the indirect effects may not show up so prominently as in a real and outside future climate. Similar interpretation problems have been met in the evaluation of air pollution research.

RESPONSE AT THE LEVEL OF STOMATA

Stomatal Density

The plant leaves lose water primarily by evaporation through the stomata. The stomatal density depends upon plant species, and can be related to the plant-ecotype (between 300 and 800 stomata/mm correlated the decrease in the stomatal density over time, observed in herbarium leaves collected over the last centuries, with the rising CO_2 concentrations and concluded from the shift in 13C that the water-use efficiency concomitantly has improved. The nitrogen content in the leaves had dropped, in line with most data from elevated CO_2 experiments.

Experimentally, an increase in [CO_2] up to about 310 l/l decreased the stomatal density, but sometimes no effect is found above this [CO_2]. This matter is still under dispute although such a correlation has also been confirmed for palaeo-records. Among species large differences in response of stomatal density to elevated [CO_2] seem to exist. Experiments with a range of CO_2 concentrations showed an increase of stomatal density of rice and bean leaves, with a differential effect at abaxial (increasing) and adaxial sides.

At subambient CO_2 concentrations stomatal density dropped. This is in contrast with the findings of Oberbauer *et al.*) for tropical trees. The relative effect of changes in stomatal density and decreasing stomatal aperture at elevated [CO_2] for the water relations, however, have not been evaluated. A gradual change in [CO_2] over the next century may lead to a natural selection that favours cultivars having a lower stomatal density, especially for water-limited growth conditions. It should, however, be realised that other environmental factors like salt stress can also modify stomatal density. Whatever the net effect, the resulting stomatal conductance is primarily determined by stomatal functioning, and much less by density.

Stomatal Functioning

In the pathway from the stomatal cavity to the leaf surface, and from ambient air to the photosynthetic machinery in the mesophyll, the stomata are a major resistance for gas transport between the leaf and the surrounding air.

A change in gas exchange resistance of the stomatal pores therefore affects the entrance of CO_2 and even more the exit of water vapour. The opening status of the stomata is a compromise between water loss and uptake of CO_2 from ambient air. In line with this proposition, stomatal response to elevated concentrations of CO_2 (Ca) in general is reflected in partial stomatal closure. The mechanism behind this stomatal closure is not yet clear. The observations are in line with the notion that plants tend to regulate the internal CO_2 concentration (Ci) in the substomatal cavity such that for a given water vapour deficit there is a constant ratio (Ci/Ca) with the atmospheric concentration.

Such a regulation would directly lead to the partial closure at elevated CO_2, as observed in many studies using porometers. Jackson *et al.* measured photosynthesis and water relations of native grassland species and calculated Ci/Ca for C_3 and C_4 plants. They confirmed the conservation of the value, with only a small (not significant) tendency to rise with increasing [CO_2]. The rate of photosynthesis and therefore the required supply of carbon dioxide is directly coupled to light intensity.

In line with this, the conductance of stomata is also highly correlated to light. This relation can be modified by environmental conditions like drought or stress by air pollution. The ratio Ci/Ca in stationary conditions is about two-thirds for C_3 but about one-third for C_4 plants. The lower value for C_4 plants reflects the higher affinity for CO_2 of the C_4. photosynthetic pathway, and reflects the more efficient water use for these plants.

Acclimation of Stomatal Movement

Little is known about the acclimation of the stomatal movement to long-term exposure to elevated [CO_2]. Some studies suggest that the lower stomatal conductance not only persists over extended exposure periods to elevated [CO_2], but is also conserved after subsequent lowering of the [CO_2]. For glasshouse horticulture it was reported that the stomatal conductance of high-[CO_2] tomato plants was less sensitive to short-term [CO_2] fluctuations, and higher than that of plants grown at ambient concentrations and measured at double present [CO_2]. This may indicate that the sensitivity of stomatal movement of high [CO_2] plants to changes in [CO_2] is reduced, but still a lower stomatal conductivity exists when compared to leaves growing and measured in present [CO_2] conditions.

EFFECTS AT THE LEVEL OF THE LEAF

The partial closure of stomata is reflected in the reduced conductance at the leaf level. At a CO2 concentration double the present, conductance is reduced by 30-40 per cent, although large differences among species exist.

Leaf Temperature and Vapour Pressure Deficit

Water loss by transpiration is not only affected by the conductivity of the stomata, but also by the driving forces for exchange of the water vapour from the leaf surface to the surrounding atmosphere. Therefore the gradient in partial pressure of the water vapour at the leaf surface is also of importance. All other factors being equal, the existing vapour pressure deficit (VPD) between stomatal cavity and surrounding air, the boundary layer, will increase at a reduced transpiration rate, and feed back to stimulate transpiration.

The reduced transpiration will cool the leafless, and consequently a rise in temperature in the stomatal cavity and at the leaf surface may occur. Thus, in addition to the global greenhouse effect on air temperature, the temperature at the leaf surface may rise by 0.5 to 1.5 °C. In their FACE experiments (Free Air Carbon dioxide Enrichment) in Arizona. measured an average rise in canopy temperature of 0.56°C over the growing season. Such a higher leaf temperature may also have important consequences for the longevity and photosynthetic capacity of the individual leaves and at the canopy level, as ageing may be accelerated. It is presently not known how different crops respond to the increase in leaf and canopy temperature.

The mostly negative feedback on water vapour exchange will at least partly counteract the effect of the reduced stomatal conductance even without rise in ambient temperature at the global scale. A rise of ambient growing season temperatures due to the greenhouse effect will also tend to stimulate transpiration. Depending on changes in air humidity it may affect VPD at the leaf surface determining the fluxes of sensible and latent heat and thereby the energy balance. Increased water availability may, however, offset the relative warming at the leaf surface, especially if, through preceding water saving and lower instantaneous requirements, partial stomatal closure alleviates water stress during periods of high water requirements in the diurnal sequence.

8

Advances in Crop Production Technologies

ADVANCED AGRICULTURAL TECHNOLOGIES IN CROP

Sometimes you're out in the middle of nowhere and need to make an unexpected repair, but you don't have the right tools. Not a problem for a farmer. In a pinch you can make something else work for you 9 times out of 10.

Farmers are also great at adopting new technology. Nowadays tools on many farms and ranches are very sophisticated. From tractors that drive themselves to cows that text, the modern farmer's toolbox contains a host of cool toys.

Here are 12 of the most advanced agricultural technologies employed today:

TRACTORS ON AUTOPILOT

Thanks to GPS tractors, combines, sprayers and more can accurately drive themselves through the field. After the user has told the onboard computer system how wide a path a given piece of equipment will cover he will drive a short distance setting A & B points to make a line. Then the GPS system will have a track to follow and it extrapolates that line into parallel lines set apart by the width of the tool in use.

These systems are capable of tracking curved lines as well. The tracking system is tied to the tractor's steering, automatically keeping it on track freeing the operator from driving. This allows the operator to keep a closer eye on other things. Guidance is great for tillage because it removes human error from overlap, saving fuel and equipment hours. Trust me when I tell you that once you starting auto tracking, you'll never go back manual steering.

SWATH CONTROL AND VARIABLE RATE TECHNOLOGY

Building on GPS technology are swath control and VRT. This is where guidance really begins to show a return on investment. Swath control is just what it sounds like. The farmer is controlling the size of the swath a given piece of equipment takes through the field. This video is a great visual representation of how swath control works.

The savings come from using less inputs like seed, fertilizer, herbicides, etc. Since the size and shapes of fields are irregular you are bound to overlap to some extent in every application. Thanks to GPS mapping the equipment in the field already knows where it has been. Swath control shuts off sections of the applicator as it enters the overlap area, saving the farmer from applying twice the inputs on the same piece of ground.

VRT works in a similar fashion. Based on production history and soil tests a farmer can build a prescription GPS map for an input. By knowing what areas of a field are most and least productive the application rate of an input like fertilizer can be tailored to increase or decrease automatically at the appropriate time. This is a big benefit for farms. Instead of applying a set rate of fertilizer over the entire field (many times a high rate to help those low producing areas) an operator can now apply a rate most effective for a particular section of ground.

YOUR TRACTOR IS CALLING

Telematics is being touted as the next big thing in ag. This technology allows equipment to talk to farmers, equipment dealers, and even other equipment. Imagine you have a problem in the field and have to stop working. With telematics your dealer can access the onboard diagnostic system of your tractor. Depending on the problem they might be able to fix your equipment right from dealer. No waiting on a mechanic to drive out to wherever you might be. You're back to work, and the dealer saved a trip too. Farmers will be able to keep track of what field equipment is in, fuel consumption, operating hours, and much more. Personally I've noticed on our farm as we become more technologically advanced our downtime is often caused by electrical, software, or hardware problems as opposed to mechanical.

Tractors can even communicate between themselves. The best example is a combine and a grain cart. Grain carts pull up next to harvesting equipment so the harvester can unload on the move without stopping to unload. Telematics can tell the grain cart operator when a combine is filling up with grain. Even better if one cart is chasing two combines. The driver knows which machine needs unloading first. The latest innovation has the combine operator actually taking wireless control of the tractor pulling the grain cart as it pulls alongside, giving him the ability to shuttle the cart forward and back to more easily fill the cart.

YOUR COW IS CALLING TOO

And it's not saying "Moo!" Collars developed for livestock are helping producers keep track of their herds. Sensors in the collar send information to a rancher's smartphone giving the rancher a heads up on where a cow might be, or maybe she's in some sort of distress, or maybe just in the mood for some mating. I suppose you could say it's kind of like telematics for cows!

RFID tags are also a handy device for livestock management. The information kept on a tag helps producers keep track of individual animals, speeding up and making record keeping more precise. I recently read about RFID tags placed in to hay as it is baled. Data such as moisture and weight can be stored in the tag to be scanned later.

IRRIGATE VIA SMARTPHONE

Mobile tech is playing a big role in monitoring and controlling crop irrigation systems. With the right equipment a farmer can control his irrigation systems from a phone or computer instead of driving to each field. Moisture sensors in the ground are able to communicate information about the level of moisture present at certain depths in the soil. This increased flexibility allows for more precise control of water and other inputs like fertilizer that are applied by irrigation pivots. Farmers can also combine this with other tech like VRT mentioned earlier to control the rate of water applied. It's all about more effective and efficient use of resources. I read a magazine article recently where a farmer stated that because of these technologies his total water usage is now less than what he was wasting previously.

SENSING HOW YOUR CROP IS FEELING

Crop sensors. This is taking variable rate technology to the next level. Instead of making a prescription fertilizer map for a field before you go out to apply it, crop sensors tell application equipment how much to apply in real time. Optical sensors are able to see how much fertilizer a plant may need based on the amount of light reflected back to the sensor. I haven't seen one of these systems in operation yet, but I'm keeping a close eye on them. It's fairly new and pretty expensive, but I see huge potential here. Crop sensors are going to help farmers apply fertilizer in a very effective manner, maximizing uptake and reducing potential leaching and run-off into ground water.

FIELD DOCUMENTATION

Because of onboard monitors and GPS the ability to document yields, application rates, and tillage practices is becoming easier and more precise every year. In fact farmers are getting to the point where they have so much good data on hand that it can be overwhelming to figure out what to do with all of it.

And of course, every farmer's favorite form of documentation is the yield map. It sums up a year's worth of planning and hard work on a piece of colourful paper. As harvesting equipments rolls through the field it calculates yield and moisture as it goes tying it in with GPS coordinates. When finished a map of the field is printed. These maps are often called heat maps. I liken then to weather radar maps. Each colour on the map relates to a certain yield range. Now the farmer can see what varieties had the best, worst, or most consistent yield over varying conditions. Maps like this can tell a farmer how well a field's drainage system is working. (In case you weren't aware, there is a massive network of private and public drainage at work underground on farms.) All this data allows for better agronomic decisions in the future.

IMPROVED CROP PRODUCTION TECHNOLOGIES

Agricultural statistics of Lesotho include data over a 15 year period on the yield of the major crops on a District basis 4. These data have been analysed and elaborated to explore the implications for investment decisions in farming.

No significant technological change has taken place in the Lesotho mountain areas in the last 15 years. Production is by small farmers, almost exclusively for home consumption. The annual yields of all crops fluctuate widely, so that the average over the period is not statistically significant. Nor can any statistically significant trend line be detected. All crop yields generally fluctuate in the same direction in the same year 5. Yield fluctuations are due mostly to different climatic factors: rainfall, hail, and frost: particularly time and duration of rainfall, time and intensity of hail, time and duration of frost.

Yield data refer to quantities produced related to area sown, as distinct to area harvested. Lesotho statistics also show the percentage of area sown to different crops which is not harvested (an indication of total crop failures). These are not published on a District basis, but refer more broadly to lowland, foothill and mountain areas. The latter information, covering a 19 years period has not been separately included in the analysis 6, since its impact is reflected in the yield data.

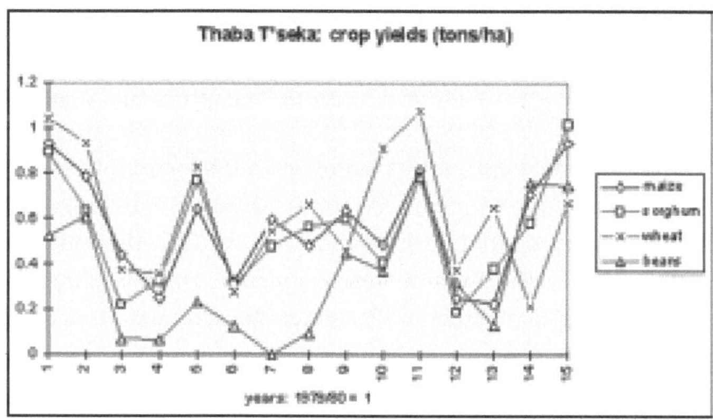

METHODOLOGY

Investment decisions by farmers, particularly decisions regarding the switch to so-called improved technologies, would be expected to depend on their perception of the risks involved relative to the possible benefits. This involves expectations about three uncertainty factors: (i) the average yield increase that the switch would bring about; (ii) the extent of fluctuations in crop yield and in returns to labour that would result from the adoption of improved technology compared to continuing with the traditional technology; and (iii) the crop prices. As against these uncertainties, the differential costs which would be incurred are reasonably firm.

To deal with the first uncertainty, several assumptions about farmers' perceptions of the expected land productivity increases have been tested. Since the statistical evidence is that average yield data are not significant, those assumptions have been combined with an analysis of the second uncertainty factor. The basic assumption in dealing with the latter is that farmers' expectations about the future fluctuations of crop yields will be influenced by the pattern of past crop performance in their area. It must be emphasized that the models are behavioural models and are not designed to predict precisely what will happen to crop yields in the future, but to assess possible farmers' responses to proposals concerning the adoption of new technologies.

In order to do that, stochastic efficiency analysis has been applied. In this procedure, the stochastic dominance concept is used as a measure of the probability of obtaining better results from one technology vs. another, given the probability distribution of a set of outcomes resulting from the application of either one of the two. Stochastic efficiency analysis requires the estimate of the cumulative distribution function (CDF) of the probabilities of a selected index of profitability. In a small farmer, food security oriented setting, gross returns to labour (revenue less cash costs) are an adequate indication of crop profitability.

To construct the CDFs, the yield data series have been arranged in ascending order, and correlated to the probability of occurrence, such probability being measured at 1/15 for each value of the series which are made up of 15 observations. From these p values, a cumulative probability series has been calculated.

The data have been applied to two crop budget models: one represents the traditional technique applied by the farmers (result of the SEPSS investigation), and the other the modern "low input" improved technology recommended by the GOL. Two series of gross returns to labour have been calculated, each having 15 values. The series are constructed on the basis of the corresponding crop budgets, keeping cash costs constant and applying the pattern of yield outcomes to crop prices to obtain a series of revenues. The difference between the revenue series and the fixed cost has been associated with the cumulative probabilities series.

Two CDF of returns to labour from maize farming. Given the way that data have been arranged in the graphs, each point of the CDF curve lying above the other has stochastic dominance, *i.e.* results in higher returns for the same probability value. At each point on the CDF the probability that returns to labour (plotted on the y axis) would be more than the corresponding cumulative probability value plotted on the x axis is = (1 - x). Thus the probability of better returns than the maximum value is (1 - 100 per cent = 0); the probability of better returns than the minimum value is (1 - 7 per cent = 93 per cent). In the case of maize, for example, the probability of exceeding a "modal" value of M300 (corresponding to about 900 kg/ha with the improved technology) is (1 - 43 per cent = 57 per cent); and so on.

The model uses a percentage yield increase factor, applied as a constant to all historical yield data, to estimate the impact of the improved practice. This procedure switches the yield profile of the improved technology upward by the same proportion in each point of the curve. The yield increase factor can be changed (as well as the crop price) to simulate the impact of different assumptions.

The above condition, and the assumption of constant production costs, may seem an oversimplification. In reality, with adverse dry farming seasonal conditions, production costs such as harvesting and fertilizers application will be less, but the yield of more sophisticated crop varieties may well be also considerably lower, the net impact on returns to labour being difficult to estimate. With very good conditions, harvesting and perhaps weeding costs will be more, and yields may well be proportionally higher with improved technologies. Such differences would be important if one was building models that would attempt to predict precisely what would happen to returns to labour. In our case, however, we are trying to identify the general conditions which would persuade a farmer to adopt or reject a new technology, and this would not be influenced much by relatively marginal adjustments.

Comparison of the two CDFs (the one corresponding to the traditional technology and the one corresponding to the improved technology) gives an indication of the stochastic dominance of one technology over the other, *i.e.* of their relative stochastic efficiency. The sum of the differentials (returns to labour with traditional technology less returns to labour with improved technology) over the entire range of observations, or 1/15 of that value on a per annum basis, gives a first indication of the preferred technology. Negative values of this sum suggest preference for the traditional technology.

In the case of stochastic dominance of the first degree, the technological preference is straightforward.

In the case of stochastic dominance of the second degree (the two curves cross somewhere) the problem is more complex. Other indices have been worked out to explore these situations. A relevant factor in decision making of small farmers would be the size of the negative values resulting from the application of one technology vs. the other, reflecting the high liquidity preference of these producers (fear of losing money). When the sum of the differential is = 0, the two technologies may be indifferent, but if the size of the sum of possible negative values is larger for one technology, the other would be normally preferred. By assuming different values of the yield increase factor, it is possible to work out the relationship between the percentage increase in yields and the gross return to labour resulting from the stochastic efficiency profiles.

From this, the minimum increase required to bring the two technologies onto an indifference level can be read, and reading it in conjunction with the CDF graphs, suggests what the minimum yield required to induce farmers to make the shift might be.

Finally, by changing the price assumption, the impact of price changes on the preference for new technology adoption has been explored.

RESULTS

Maize. Maize is by far the most important crop, grown over about 70 per cent of the cropped area in the District. The 15 year AVG maize yield (0.563 tons/ha) falls between a MIN of 0.222 of and a MAX of 0.913.

The maize crop budgets compare two farms equipped with their own oxen, plough and harrow. The annual cost of the equipment is estimated by applying a capital recovery factor at 10 per cent interest over the expected life of the equipment and amortized over the number of hectares tilled (either farmers' own, or through contract work).

The comparison involves essentially the use of hybrid seeds and fertilizers in the improved technology. In a first scenario, the price of maize is assumed to be M1,000/ton. This price reflects market conditions in the area at the time of the SEPSS, and compares with a Government support price of M 610/ton (Maseru), and with a price prevailing in RSA of M 400/ton. In a first approximation, the latter would represent the border value to be used for economic evaluation.

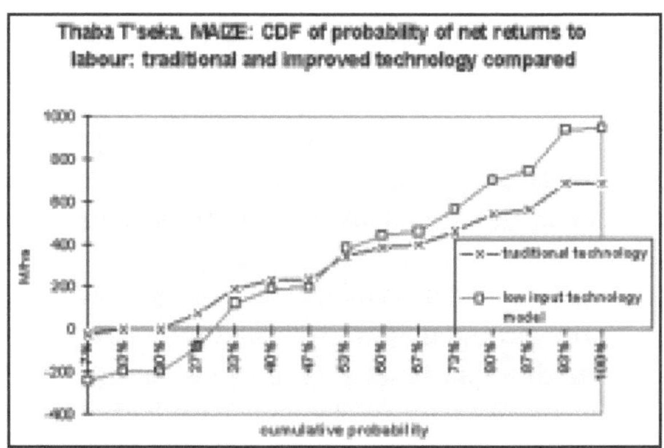

Fig. (maize price = M 1,000/t, yield increase factor 1.66)

The expectation of a price of M1,000/ton and of a yield increase of 66 per cent the improved technology, the traditional and the improved technologies are on an indifference position if total expected returns are compared. However, the sum of probable negative returns is much less with the traditional technology, which suggests that the improved package would not be adopted.

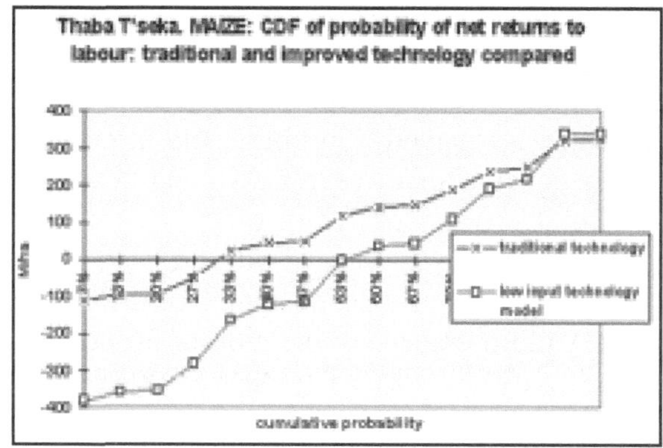

Fig. bis (maize price = M 610/t, yield increase factor 1.66)

The results are much worse if the GOL support price for maize (M 610/ ton) is applied to calculate returns to labour. In this case the traditional technology shows a stochastic dominance of first degree: no rational farmer would adopt the improved technology with such a price expectation. It is interesting to explore the effect of price liberalization. If the RSA border equivalent price is expected, improved technology has even less chance, and the probability of traditional technology showing positive returns to labour is only 25 per cent. The implications are that there is no comparative advantage for the production of maize in the area, no matter which technology is selected, and that the relative advantage of traditional technologies increases as output prices fall.

The conclusion that a package involving use of fertilizers, chemicals and hybrid seeds would not be viable if a highly variable climate prevents the attainment of expected yield increases commensurate with the genetic potential of the material planted can of course be reached intuitively. What cannot be intuitively arrived at is the relationship between the expectation of an average yield increase, the price of the crop, and the likelihood of adoption of a technology which implies a given increase in cash costs.

Conclusion on maize

Farmers may well continue to grow maize in the Mountain Areas of Lesotho, but they would be ill-advised to try and improve their technology by the route suggested (hybrid seeds and fertilizers). While emphasis on hybrid varieties seems clearly ill placed, there may be a market for composite seeds, which would tilt the scale with the expectation of high prices. Such an expectation could indeed reflect the farmers' preference for "their own maize", and may explain the persistence of the crop in an unsuitable area. The situation calls for launching a programme of applied research aimed at reducing the sensitivity of crop yields to climatic conditions rather than at increasing the crop yield potential.

Wheat

Wheat is the second most important crop in the area, being grown over 15-20 per cent of the area cropped. (Sorghum, the other main cereal grown, accounts for 5-10 per cent). The 15 year AVG yield of wheat (0.624 tons/ha) falls between a MIN of 0.210 and a MAX of 1.075.

A similar analysis applied to the case of wheat demonstrates the sensitivity of the stochastic efficiency of improved low input technology to the price of the wheat crop. Given a yield increase expectation of 50 per cent, and the high price prevailing at the time of the SEPSS in the area (M 1250/ton), the improved technology essentially shows a stochastic dominance of the first degree, *i.e.* preferable in (almost) all points of the CDF.

Applying the GOL support price of wheat (M 650/ton) a situation of stochastic dominance of second degree prevails. A 50 per cent yield increase would result in the two curves showing an apparent indifference between the two technologies, except that negative returns from the traditional technologies are less than with the improved technology.

Conclusion on wheat. Wheat yields seem to be declining marginally in Thaba T'seka.

Despite the agronomic advantage that the crop ought to have in the climatic conditions of the area, wheat is not a preferred crop in Thaba T'seka. Farmers are unlikely to be enthusiastic about the improved package, unless it can be shown that yield would increase significantly more than 50 per cent. Spending public money to extend the present improved wheat technology is hard to justify. Research efforts should concentrate on both higher yield potential and better resistance to adverse local conditions.

Beans

Pulses are grown over 5-10 per cent of the area cropped. The 15 year AVG bean yield (0.35 tons/ha) falls between a MIN of 0 and a MAX of 0.788.

The case of beans, selected as a proxy for a number of pulses, shows a definite stochastic dominance of the improved technology package at the market price of M3,150/ton and a yield increase expectation of 66 per cent.

However the high risk of switching to the new technology, indicated by the size of possible negative results (twice that of traditional technology) is clearly evidenced.

Conclusions on beans. The improved technology is likely to be of interest, particularly if the yield increase expectation is more than 66 per cent, which is possible with legumes (sensitivity to fertilizers and, above all, pest control applications).

OTHER LESOTHO DISTRICTS

The same analysis has been carried out for a variety of crops from data relating to six other Districts of Lesotho. Similar results have been obtained in most cases. Only in Leribe, where natural conditions are distinctly better, the recommended improved technologies show stochastic efficiency dominance over traditional practices. This confirms the conclusion that conventional technological solutions, often imported from other areas in which different agro-ecological conditions and different farming systems prevail, do not provide adequate responses to the deep routed problems of agricultural production in the country.

Crop		Thaba Tseka	Qacha's Nek	Mokhot-long	Mafe-teng	Quthing	Leribe
Maize	*Probability of*						
	Yield <average yield	47%	60%	53%	67%	60%	53%
	Yield <50% of average	40%	13%	0%	20%	13%	0%
Wheat	*Probability of*						
	Yield <average yield	47%	40%	60%	47%	66%	33%
	Yield <50% of average	13%	20%	0%	40%	67%	0%
Sorghum	*Probability of*						
	Yield <average yield	53%	40%	60%	67%	47%	67%
	Yield <50% of average	13%	13%	20%	13%	7%	20%
Beans	*Probability of*						
	Yield <average yield	53%	73%	73%	73%	60%	53%
	Yield <50% of average	40%	40%	40%	27%	47%	27%

Fig. Summary of farming risk in six Lesotho Districts

Only Leribe District offers reasonable chances of farmers adopting improved technologies based on chemical inputs and high yield crop varieties. Quting, though a lowland District, offers worse opportunities that in the Mountains: it is one of the driest areas of the country. Of the three mountain Districts, Mokhotlong seem to offer less risky farming opportunities, particularly to maize growers.

The farming risk in the six Districts, measured by the probability of crop yields being less than the average yield of the 15 years time series, and less than 50 per cent of that average. The stochastic dominance profiles of traditional vs. "improved" practices for different crops to the farmgate price of the crop.

IMPLICATIONS FOR PROJECT DESIGN

Cropping in the mountain areas of Lesotho is to a fairly large extent on terraced land: some terraces, however, show signs of degradation. Ox drawn equipment is almost universally used for ploughing, harrowing and, in the case of maize sorghum and pulses, seeding as well.

Small farmer households face a situation of very high risk, with almost 50 per cent chance that their crop yield will turn out below average. The chances of total crop failure are between 25 per cent of the sown area for wheat and sorghum and 31 per cent for maize. Correcting the situation by way of investment in irrigation is out of the question for food crops (topography involves prohibitive costs for other than micro-schemes). There may be some scope for high value crop production (fruits, flowers). These crops require high technical standards, good quality control, sophisticated packing, and ready access to transport infrastructure, to ensure successful marketing: the comparative advantage in these lines would almost certainly be in the lower land areas of the country.

The moment, the achievement of economic returns from agricultural development in the mountain areas, adequate to justify the cost of externally financed projects appraised in accordance with established procedures, cannot be expected. A preliminary estimate shows DRCs well above parity of international market prices when improved agricultural practices are simulated, and for traditional technologies as well. This confirms the generally held view that the resources of the mountain areas are inadequate to support the population presently settled there, and that long term solutions have to be found outside the agricultural sector, possibly involving substantial out-migration.

Until the economy of Lesotho has developed sufficient job opportunities to absorb a large share of the mountain area population in non-agricultural activities, food security will remain the main priority of the people settled in those areas, and a priority objective of Government policy, with a view to holding people there at an acceptable standard of survival. Farmers will aim to reduce the cash cost of cropping as much as possible, and therefore are likely to adopt technologies which minimize external input requirements, maximize in situ capture of moisture, and increase crop resilience to moisture stress and to other climatic hazards.

Several considerations relevant to the design of agricultural development policy for the Mountain Areas of Lesotho can be derived from the analysis:

1. In terms of the value of crop produced for human nutrition, there is a 47 per cent chance that the average farm land available per caput (0.3 ha) will supply less than 2,500 calories per day with the traditional technology. Available improved technologies for cereals which could obtain a 50 per cent yield increase would reduce that probability to about 15 per cent, but the financial risk involved for the farmers prevents this solution from being widely accepted. Potatoes, a crop recently introduced in the area, appear to offer the potential to bridge the gap. The nutritional production implications of farming this crop over 10 per cent of the land cropped with traditional technologies, at the expense of maize, wheat and sorghum, have been simulated, applying to potatoes the frequency distribution of wheat yields (annual yield data for potatoes are not available). This scenario would reduce the probability of inadequate calorie production to 20 per cent, and increase overall food availability to a significant extent.
2. Agronomic research ought to concentrate on crop varieties which

would be more resilient to the climatic hazards of the zone. A thorough review of the work done so far for the area, and of the potential offered by research in similar areas all over the world must be undertaken. While basic research is out of the question in Lesotho for financial and other reasons, contracting out research programmes to institutions abroad is advisable, and applied research in the area on possible promising improved varieties must be part of the development measures proposed on a priority basis.

3. While most crops in the area are grown as pure stands in the farming system, the Machobane developed technique for continuous intercropping of several crops offers interesting advantages, particularly with respect to the risk reduction objective. This technology is beginning to spread in the area, but is not yet the object of official support. Lesotho research organizations ought to monitor solutions found by Machobane farmers in adapting the technology to local conditions, as well as the constraints faced. Providing the link between research and Machobane farmers is a GOL function which deserves top priority.
4. Investing public money on extension of presently available technical solutions by expanding the Village Level Workers network is not justified, except possibly for potatoes. Efforts should be concentrated on providing an adequate linkage between research and farmer innovators by improving the professional capacity for understanding local production problems at the Subject Matter Specialist level, provided that a close dialogue with the research stations can be effectively established.
5. Promoting private sector investment in seed multiplication (co-operatives, groups, or individuals) may be justified in the case of composite maize seeds, as well as potatoes.

FURTHER ANALYSIS OF THE CASE OF MAIZE

Let us go back to the analysis of the case of maize and explore the possibility of establishing an empirically derived relationship between the yield variance due to non-predictable factors, the level of the crop price, and the stochastic efficiency of a new technology relative to that of the traditional technology. This would enable the analyst to respond to the following question: what is the level of productivity increase that a farmer would consider sufficient to induce him to adopt a new technology, given the incremental cost to be incurred, the crop prices that may confront him, and the fluctuations in yields which he is accustomed to experience in his area?

To explore that relationship it is assumed that a reasonably prudent innovative farmer would adopt a new technology if the chances of negative returns to labour were no greater than those from currently applied technology. This is a stringent condition, near to stochastic dominance of first degree, but one that would correspond reasonably well to what is known about the "risk aversion" of agricultural producers, particularly those who have a high liquidity preference (fear to lose money) as most small farmers do.

This compares the average yield increase factor expected from the recommended improved maize technology which brings about (i) a target 50 per cent increase of returns to labour, and (ii) the stochastic dominance of the improved technology over the traditional one. The stochastic dominance condition of the improved technology is defined as in the previous paragraph.

With a price of M1,000/ton, an average yield increase of about 90 per cent (yield up to 1.14 t/ha) would generate 50 per cent higher returns to labour; but the technology would have stochastic dominance (*i.e.* would be accepted) only if the expectation was to achieve an average yield 160 per cent higher than with traditional technology (1.9 t). With a lower price (M 600/ton) the corresponding values are 116 per cent (1.3 t) and 260 per cent (about 2.1 t), respectively. The lower the price the larger the spread: with the price prevailing in the RSA (M 400/t) the graph is meaningless since both technologies show negative returns.

The two curves have a different shape, the higher one reflecting the shape of the CDFs from which it is derived. The spread between the two curves depends on the yield fluctuation data upon which the stochastic efficiency curve is built. The less acute the fluctuations, the more the two curves will tend to coincide, and viceversa. This result has wider implications. In the Lesotho context, it suggests that it may be more important to develop technologies which would tend to reduce yield fluctuations, rather than increase the absolute productive potential.

From the above data has been constructed. The simplest way to grasp the meaning is to think of the sight of a gun, which must be raised higher, the further away is the target, and the less powerful is the charge in the cartridge. As in a ballistic table, the graph shows how much higher one needs to set the sight in order to hit the target. Naturally, as with all guns, the sight cannot be raised to hit beyond the range of the gun.

For example, that, in the context of the Thaba T'seka experience, with an expected maize price of M 800/ton, a dominant (acceptable) technology ought to obtain 45 per cent better yields than the one which would obtain, on average, 50 per cent more returns to labour than the traditional technology. If the higher target is technically not feasible, the technology would be rejected.

When embarking on a new venture, a reasonably prudent decision maker would attach more weight to the risk of obtaining larger negative results than to the opportunity of reaching larger positive results. This can intuitively be accepted as a feature of the basic psychology of most decision makers. Most agricultural producers operating under conditions of fluctuating outcomes, to be persuaded to change from a well known technology to an expected superior one, would require the expectation of obtaining average returns larger than those which they would accept in the absence of fluctuations, with a view to offsetting the larger losses expected to be incurred when adverse circumstances prevail.

Can a practical method be devised to assess how much larger that expectation would be? A closer look would suggest that a general rule might be at work, applicable not only in the extreme circumstances of Lesotho's mountain areas, but in all situations in which fluctuating economic results have been usually experienced in farming, and would be expected from adopting a new technology as well. The rule can be formulated as follows:

Whenever the returns of a technology innovation are expected to fluctuate as a result of fluctuations of independent variables affecting those results:

(i) The wider the fluctuations of the independent variables, the larger would be the difference between the expected average result which would induce the technology change, and the expected average result set as a target, and

(ii) The lower the expected price of the product, the larger would be the difference between the expected average result which will induce the technology change, and the one expected to achieve the target on average.

The rule would apply to all technology changes which require an increase in cash costs per unit of land, and would operate more forcefully, the more the risk aversion of the producers would be close to the criterion set in paragraphs 8 and 41.

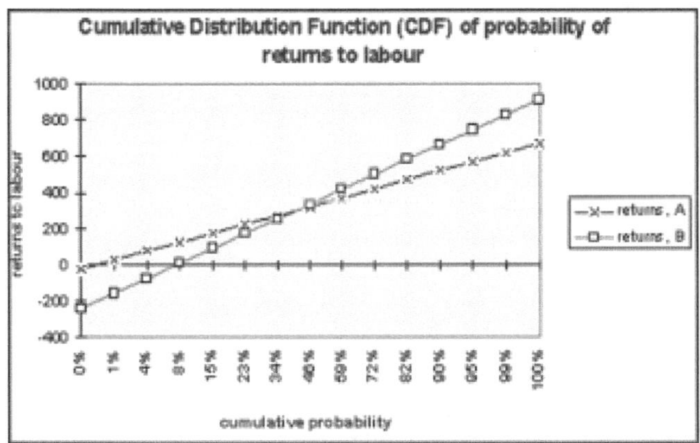

Fig. Maize yields distributed triangularly, maize price fixed.

In many countries, the actual production performance of many Agriculture and Rural Development Projects has fallen short of expectations. It is suggested that some of the disappointing results might have been predicted if the rules stated above had been taken into account in project evaluation. A method to measure the spread between the expected incremental results required to achieve project targets on average, and those required to induce the technology switch.

In many countries/project areas, however, sufficient data to perform the full analysis might not be available. A short-cut method has accordingly been explored, with a view to providing a "rough and ready" indicator of the requirements for technological innovations to be accepted in a small farmer environment.

The analysis has been repeated for the maize crop, based on a value for the expected most likely yield, and for the minimum and maximum yield. The values used are the same as in the previous analysis, *i.e.* the actual minimum and maximum yields recorded, and the yield increase factor of the improved technology. The three yield points have been used to estimate the probability frequencies distribution of a series of equally spaced yield data, comprised between the minimum and the maximum, assuming such distribution to be triangular. The results are remarkably similar to those obtained using the full 15 years yield statistics. The curve A represents the traditional technology, the curve B represents the improved technology.

The triangular distribution method can be used equally well in cases when the yields fluctuations are not very important, but significant price fluctuations may occur. The simulation of the maize technology analysis obtained by assuming a fixed yield increase factor of 1.9 and fluctuating price distribution with MIN=400 M/ton, Mode=610 and MAX=1000 spread over 15 intervals.

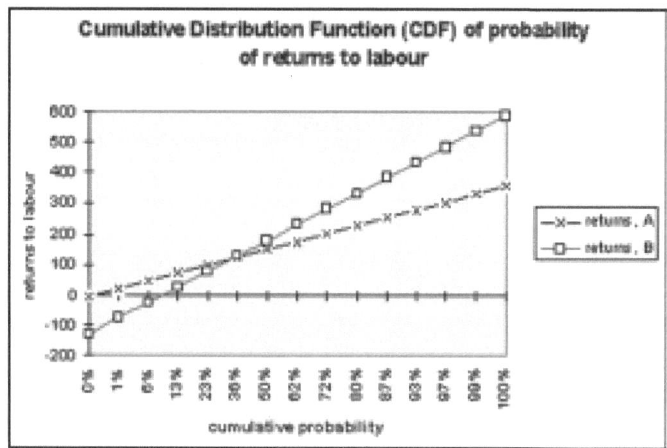

Fig. Price fluctuations distributed triangularly, maize yield fixed.

The analysis of the Lesotho case confirms Anderson and Dillon's suggestion that the triangular distribution assumption can be used to construct the CDF of the frequencies of an independent variable in the absence of adequate time series of actual data which are acceptable approximations for the purposes of risk analysis. The basic parameters can be "guesstimated" by experienced agronomists, if actual observations are not available from the field, or of very limited coverage. These can be inferred from climatic data and/or other agronomic considerations, and by asking farmers about their past experience with crop yields and price fluctuations in the context of RRA/PRA investigations. The CDFs calculated on the basis of the triangular distribution assumption.

The application of the methodology will save time spent at project appraisal on performing cumbersome hypothetical "sensitivity analyses", while providing a logical framework for simulating farmers' likely reactions, and avoiding over-optimistic assumptions about technology adoption rates.

Simulations derived from triangular distributions of the frequencies of the main variables which determine the selected indcx of profitability would give us an idea of the extent to which the performance of a new technology should exceed the performance of an established technology, for the former to be worth placing in an extension package. The analysis presented in this note suggests, for example, that, for a technological package to be accepted by small producers, if price and yields fluctuate between two extremes, one being about 2 times higher than the other, and those fluctuations are not correlated with one another, the expected average yield ought to be some 30 per cent to 40 per cent higher than the target yield which would in theory obtain adequate incremental returns to labour. Expected fluctuations of only one of the variables would obviously have much less, but always a noticeable impact.

9

Agricultural Economy of Oilseed Crops

OILSEED CROPS OILSEED CROPS

Important points for obtaining maximum Yields:

1. Using the improved variety recommended for the tract.
2. Using good and bold seeds for sowing.
3. Treating seed with fungicide, *e.g.* Captan or Thiriin, at the rate of 4 g 1 kg of seed.

SAFFLOWER

The three centers of origin, as suggested by Vavilov for safflower (Carthamus tinctorius L.), are India, Afghanistan and Ethopia. Decandole was of the opinion that Arabia was the most probable centre of origin.

Economic Importance

The crop is now cultivated, primarily for its seeds which yield oil, though at one time it used to be grown for the extraction of a dye also. The seeds are edible and are eaten after roasting. Their oil content varies from 24 to 36 per cent, depending on the variety, soil, climate and other conditions. The cold-pressed oil is golden yellow and is used for culinary purposes, or for making soap. The oil obtained by dry hot distillation is black and sticky and is used only for greasing well ropes and leather goods exposed to water. Safflower oil has also good drying properties and is, therefore, used in the manufacture of paints, varnishes and linoleum. It can be mixed with white paint without any after-yellowing effects. The cake, particularly from decorticated seed, is used as a concentrated cattle feed, and that from undecorticated seed is sometimes used as a manure.

Botanical Description

The safflower plant is highly branched, herbaceous, thistle-like annual, varying in height from 30 to 160 cm. It has many branches each terminating in a flower. The inflorescence is broad, flat or slightly curved and densely bristled owing to the presence of numerous floral tracts. The flowers are yellow to red, containing 20 individual ficrets, each of which produces a seed.

Distribution Area and Production

The important safflower growing countries, besides India, are the USA, Mexico, Ethopia, Spain, the USSR. and Australia. In India, it occupies 590,000 hectares with a production of nearly 130,000 tonnes. Over 98 per cent of the area is concentrated in the states of Maharashtra (04.4 per cent), Karnataka (26.0 per cent) and Andhra Pradesh (8.0 per cent).

Climate and Soil

The crop is grown during rabi primarily as a rain fed crop, but in some areas it is raised under irrigation. At all stages of growth of the crop, excessive rainfall or humidity increases damage from fungal diseases. Water-logging due to poor drainage or prolonged rains even for relatively short periods can cause substantial fall in seed-yield. Being drought-resistant, it is cultivated on all types of soil, including sandy soil, but it thrives best on water-retentive black soils and alluvial loams. The crop is fairly resistant to saline condition.

Caltivation

Saffiower is grown mostly as a mixture with wheat, barley, gram and rabi jowar, three rows of safflower being planted after every nine, 12 or more rows of the main crop. Sometimes, it is sown as border rows surrounding the crops of wheat, rabi, jowar, etc., because being spiny, it protects the main crop against cattle trespass. As a pure crop, it is rotated with wheat, gram and rabi jowar. Occasionally, it is grown as a second crop after any quick-maturing kharif crop, *e.g.* green-gram (moong), black-gram (mash), groundnut, coriander, etc.

The number of ploughings should be as few as possible and be aimed at preserving the maximum amount of soil moisture. One or two ploughings with a country plough will be sufficient to remove the weeds and break the clods. Sowing is generally done in September and October. The seed-rate adopted varies from 5 to 12 kg per hectare, depending upon the soil fertility and the nature of the crop (pure or mixed). When grown in strips, a row spacing of 45 cm is adopted.

Plants too close tend to have thinner stems or wear a superficial root-system, with a reduced number of flowers. In general, the pure crop is given one or two weedings, combined with hoeing on the 2Oth and the 45th day after sowing. The application of 20-40 kg of N per hectare results in a substantial increase in yield. The mixed crop shares the preparatory tillage, manuring and cultivation given to the main crop. When the plants have developed the central flowering head, they are topped to promote branching, flowering and seed production.

Harvesting and Yield

The crop matures in 4 or 5 months after sowing. The ripe plants are either cut or pulled out, stacked for a few days to dry, threshed by cutting with sticks, and the resulting material is winnowed to obtain clean seeds. The average yield of the pure crop is 400-500 kg of seed per hectare, whereas that of a mixed crop is about 100 kg.

Pests and Diseases

The safflower aphid (Dactynotus carthami), thrips (FranklinieUa sulphurea) and the flower bud fly (Acanihiphilus helianthi Rossi) are the serious pests of the crop. Spraying the crop with O.1 per cent Lenition, 0.03 per cent Dimethoate and 0.07 per cent Endosulfan is recommended for controlling aphids, trips and bud fly, respectively.

Rust (Puceinia carthami), alter aria leaf-spot (Alternai carthami) and root-rot (Phytophthora drechskri) are the important diseases of safflower. The growing of resistant varieties is recommended to avoid damage from these diseases.

Rust (Puceinia carthami), alter aria leaf-spot (Alternai carthami) and root-rot (Phytophthora drechskri) are the important diseases of safflower. The growing of resistant varieties is recommended to avoid damage from these diseases.

CASTOR

Geographic Origin

According to the available literature, castor (Ricinus comrnunis L.) is indigenous to Eastern Africa and most probably originated in Ethopia.

Economic Importance

It is generally grown for its,oil-yielding seeds. The oil content of the seeds varies from 35-58 per cent in different varieties, the average being about 47 per cent. Castor-oil is being used widely for various purposes. It is used as a lubricant in high-speed engines and aeroplanes, in the manufacture of soaps, transparent paper, printing-inks, varnishes, linoleum and plasticizers. It is also used for medicinal and lighting purposes. The cake is used as a manure and plant stalks as fuel or as thatching material or for preparing paper-pulp. In the silk-producing areas, leaves are fed to the silkworms.

Botanical Description

Castor belongs to the gellus Ricinus, its a member of the Euphorbiaccae which consists a vast number of plants mostly native to the tropics. The geiut; Ricinus is considered to be momotypic and R. cwnmunis as the only species which includes many polymorphic types. The cultivated types are dwarf annuals The stem is erect, circular in section, partially, hollow, smooth, glabrous, with good branching. The stem is marked by a number of well-defined nodes, from each of which a leaf arises.

The lower internodes are shorter and their length increas/';[p-0es with the height. Leaves are alternate, large, palniate with 5-11 lobes, acuminate, margins notchedy serrate or indented. They are carried on long stout petioles. The inflorescences are borne terminally on the main and lateral branches. Flowers are large, in terminal sub-panelled racemes, monoccious, apetalous, the upper portion of the raceme being occupied by the female flowers and the lower by the male flowers. The fruit is a roundish glaucous capsule, with three projecting sides covered with tough spines or smooth, three-loculed and three-seeded. Seeds are albuminous, anatropous, broad, oval, compressed with a marked carnnele and longitudinal raphe. The testa is thin, brittle, varying in colour and mottling. The testa is the thin tegmen, covering the whitish oily endospeim containing the embryo.

Distribution Area and Production

The main castor-growing countries are Brazil, India, USSR and Argentina. India ranks first in area (44,8000 hectares) but second in production (140,000 tonnes), Brazil being the, first. Andhra Pradesh (67.2 per cent), Gujarat (12.7 per cent) Karnataka (7.1 per cent) and Orissa (5.8 per cent) account for over 90 per cent of the area and also production.

In lighter soils, the land is repeatedly ploughed in summer, as and when the rains are received, and with the onset of monsoon rains the clods are crushed by working a country plough or harrow. In clay loams, only harrowing is done with a blade-harrow two to three times to bring the soil into proper tilth.

The seed is sown either in the plough furrow or with a seed-drill or by hand-dibbling. The seed-rate used for sowing behind the plough is 12 kg per hectare. For hand-dibbling, 8 to 10 kg of seed will be sufficient per hectare. Before sowing, castor seed should be treated with Thiram 3 g/kg of seed to prcvcnt thc attack of root-rot and Altcrnaria blight. The spacing adopted generally ranges from 90-120 cm between rows and 45-60 cm between plants in the row. One or two weedings and hoeings with bullock-drawn implements are given and sometimes, the plants are earthed up. The fertilizer dose recommended E for castor is 40 N-40 P-20 K kg/ha. The entire dose of P and K and half of W is applied basally and the remaining half of W about 30 days after sowing, as and when the soil moisture is adequate.

The improved varieties cultivated at present are early maturing and take about 150 to 180 days. When one or two capsules in a bunch show signs of drying, the whole cluster is generally removed and stacked, covered and weighed in a corner of the field or in a pit. The harvesting of unripe capsules in this manner has an adverse effect on the oil content of the seed. It is preferable to collect the fruits, as and when they ripen. When the whole crop is gathered, it is dried in the sun for a few days and the threshiug is done either by beating the dried capsules with a stick or by treading them under the feet of bullocks. Winnowing is done in the usual manner.

Pests and Deseases

The castor semi-looper (Achow janaga L.) and the capsule-borer (Dicitocrocis punctiferalis) are the two most serious pests of castor. Dusting BHC 10 per cent in early stages or spraying 0.44 per cent Ekalax or 0.1 per cent Carbaryl or 0.05 per cent 1 arathion on the crop will give an effective control on these pests.

The seedling blight (Phytopiathora colocasic Racib. emend. Thomas and T.S. Ramkar) and Alternaria blight Alternaria ricini (Yoshii) Hansford cause serious losses to this crop. The Sowing of the crop in low-lying and water-logged areas should be avoided to prevent the seedling blight from appearing. For alternaria blight, seed treatment with Thiram at the rate of 3 g/kg of seed is recommended.

Yield

The average yield of rain fed castor varies from 200 to 500 kg per hectare, that of the mixed crop from 100 to 200 kg and of the irrigated crop from 500 to 800 kg. In eleven years (from 1964-65 to 1974-75), the average total yield of castor in India has risen to 350 kg per ha.

Varieties

The varieties differ in the branching habit of the plants, colour of the stem and branches (red and green), the nature of capsules (smooth or shiny), duration (early or late) and the size of seed.

The following are the principal improved varieties of castor recommended for cultivation in different states.

RAPESEED AND MUSTARD

Under the names rapeseed and mustard, several oilseeds belonging to the crucieferae are grown in India:

- Brown mustard, commonly called rai (raya or laha)—Brassica juncea (L.) Czern. and Coss
- Sarson
- Yellow sarson—B. campestris L. var. sarson Prain
- Brown sarson—B. campestris L. var. dichotoma Watt
- Toria (lahi or Maghi Labi)—B. campestris L. var. toria Duth.
- Taramira or tara (eruca sativa Mill.)

In trade, sarson, toria and Taramira are known as rapeseed, and rai as mustard.

Banarsi rai (B. nigra Koch.) which does not fall under any of the four groups is a garden crop used as spice. The cultivation of white mustard (Brassica alba syn. Sinapis alba) is no longer found in India. Rai and Yellow sarson are self-fertile and the rest of the cruciferous oilseeds, *viz.* brown sarson, toria, taramira, Banarasi rai and white mustard, are self-incompatible.

The acreage under yellow sarson is scantily (mainly in Bihar) and constantly on the decrease. In the recent past, the accreage under brown mustard is steadily on the increase (over 65 per cent) at the expense of other Brassica due to its higher production, greater to pests and diseases and moisture stress.

Origin

Brassica juncea L. (rai) was originally introduced from China into north-eastern India. From where it has extended into Afghanistan via the Punjab. Eastern Afghanistan, together with the adjoining north-western India is one of the independent centres of brown sarson (Brassica campestris var. brown sarson). Yellow sarson (B. campestris var. yellow sarson) is commonly grown in the eastern parts of India where it shows much diversity of forms. Taramira is a relatively recent introduction into India. It is believed to be a native of southern Europe and North Africa.

Economic Importance

Rapeseed and mustard yield the most important edible oil content of the seeds of different ranges from 30 to 48 per cent. In the case of white mustard, the oil content ranges from 25 to 33 per cent. The oil obtained is the main cooking medium in Northern India and can not be replaced by any other edible oil. The seed and oil are used as a condiment in the preparation of pickles and for flavouring curries and vegetables. The oilcake is mostly used as cattle feed. The leaves of young plants are used as a green vegetable. The use of mustard oil for industrial purposes is rather limited on account of its high cost.

Botanical Description

Rapseeed and mustard include annual herbs ranging in height from 0.45 to 1.75 m. Roots, in general, are long and tapering. Toria is more or less a surface-feeder and brown sarson has long roots, with a limited lateral spread, enabling its successful cultivation under drier conditions. Yellow sarson has both extensive and lateral spread. The height of the stem varies from 45 cm (in some varieties of toria) to 1.75 m (in yellow sarson). In toria and brown sarson, the branches arise at an angle of 30° to 40°. In yellow sarson, branches arise laterally at an angle about 10° to 20° and give the plant a narrow and pyramidal shape. The inflorence is coroymbose raceme. In the case of sarson the four petals are spread apart, whereas in brown sarson and toria the petals overlap or may be placed apart, depending upon the variety. The flowers bear a hypogynous syncarpous ovary. In brown sarson and toria, the ovary is bicarpellary, whereas in the case of yellow sarson, it may also be tri or tetra-carpellary.

The fruit is a siliqua. The pods are two-valved, three-valved or four-valved, depending upon the number of carpels in the ovary. The flowers begin to open from 8 a.m. and continue up to 12 noon.

Distribution, area and Production

The crop is grown both in subtropical and tropical countries. In Asia, it is chiefly grown in China, India and Pakistan. It is also grown in Europe, Canada and the USSR, but the forms of rapeseed and mustard grown there are different from those grown in India.

India occupies the first position, both with regard to acreage and production of rapeseed and mustard in the world. In India the Braddica crops occupy the second largest position after groundnut, with 3-5 million hectares, producing about 2 million tonnes of seed annually. The chief states producing them are Uttar Pradesh, Punjab, Haryana, Assam, Bihar, Madhya Pradesh, Rajasthan, West Bengal and Orissa.

Climate and Soil

The rapeseed and mustard crops are of the tropical as well as of the temperate zones and require relatively cool temperatures for satisfactory growth. In India, they are grown in the rabi season from September-October to February-March. The rapeseed and mustard crops grow well in areas having 25 to 40 cm of rainfall. Sarson and toria are preferred in low-rainfall areas, whereas raya and toria are grown in medium and high-rainfall areas respectively.

The rapeseed and mustard thrive best in light to heavy loams. Raya may be grown on all types of soil, but toria does best in loam to heavy loam. Sarson is suited to light-loam soil and taramira is mostly grown on very light soils.

Cultivation

A fine seed-bed is required to ensure good germination. In irrigated areas, the first ploughing is done with a medium sized soil-turning plough, followed by two to four ploughing with a desi plough or a cultivator. Sohaga (Planking) is given after every ploughing.

In rain fed areas, one to two ploughings with a desi plough or a cultivator, each followed by planking, may be given. Toria, in particular, requires a fairly moist seed-bed for good germination, but excessive moisture should be avoided. The rate in the case of mixed cropping depends on the proportion of the rapeseed to the main crop.

When sown pure, 5 kg of seed per ha is used for all rapeseeds and mustard. When sown mixed with other crops, the time of sowing rapeseed and mustard is governed by the sowing of the main crop. The first half of September is best for sowing toria (if wheat is to follow, it should be sown by the end of August), 25th September to 15th October for sarson, 30th September to 15th October for raya, and tarantira is sown throughout October. The seed from healthy and desirable plants, grown in isolation in the case of self-sterile form, should be used. Whenever moisture in the field is inadequate, the seed is mixed with moist soil and kept overnight. For distributing evenly, the seed is usually mixed with sand before sowing.

In mixed cropping rapeseed and mustard are sown in rows 1.8 to 2.4 metres apart across the main crop. The pure crop of rapeseed or mustard is sown at a depth of 4 to 5 cm in lines, 30 cm apart, with a drill, or with a (tube) attached to the plough. Thinning is done three weeks after sowing to maintain a plant-to-plant distance of 10 to 15 cm.

One hoeing in the case of toria in the third week after sowing and 1-2 hoeings in the case of sarson and raya are adequate. Forty kg of W per ha is optimum for all rapeseed and mustard crops in rain fed areas. Under irrigated conditions doses of 40 and 80 kg of W per ha are considered optimum for toria, sarson and rapeseed respectively. For tarantira, 20 kg of W per ha is ancient. All the fertilizers should be drilled before sowing. Among the Brassicae, raya is most responsive to irrigation, followed by yellow sarson. Two irrigations, one at flowering and the other at pod formation, result in the maximum yield in the case of toria, sarson and raya. Its flowering is most economical, if there is water scarcity. Tarantira is not irrigated.

Harvesting is done as soon as the crop begins to turn yellow. Tma, which takes 75 to 90 days to mature, is the earliest crop to be harvested. Harvesting is done with hand-sickler. Threshing is done by beating with a wooden stick the seed bearing part of the plants, taken in convenient bundles or by trampling them under the feet. Winnowing is done with by slowly dropping the threshed produce from a basket held shoulder-high. The seed after being dried in the sun is stored in gunny bags or bins.

Pets and Diseases

The most serious pests of rapeseed and mustard are the mustard aphid (Lipaphis erysinbi (Holt) syn. Rhopalosiphun pseudobrwsicae (Davis), mustard sawfly (Athalia proximo King) and cutworm (Agrotis ypsi" Poott.). To control aphidy the spraying of Methyl Demeton 0.02 per cent or limetboate 0.03 per cent two or three times, depending on the intensity is recommended. The mustard sawfly is controlled by dusting 10 per cent BHC at the rate of 25 kg per ha. Cutworms are controlled by drilling into the soil 5 per cent dust of Aldrin or Heptachlor or Chlordane at the rate of 25 kg per ha.

Alternaria blight (Alternaria brassicae Bork.) is the most widespread and destructive disease of rapeseed and mustard, whereas rust (Cystopus candidus) is serious in certain areas. Spraying with Dithane M-45 or Difolaton at the rate of 11 kg per ha is recommended for controlling these diseases. Orobanche, a phanerogamic parasite, has also been reported to be serious in Punjab and in certain parts of Uttar Flradesh. Hand-pulling of the plants of the parasite and a long rotation are recomnended.

Yield

The average yield of rapeseed and mustard is about 500 kg per ha. Under good crop management, toria, which is a short-duration crop, gives a yield of about 450 to 650 kg per ha, sarson 800 to 1,000 kg and rai 1,000 to 1,200 kg per hectare.

Oil and Protein Quality in the Brassicae

The oil of rapeseede and mustard possesses a Sizable amount of crucie acid (38 to 57 per cent), together with linolenic acid up to 4.7 to 13.0 per cent. The oleic and linolenic acid, which have nutritive value, together constitute only about 27 per cent. It is desirable to increase the quantity of oleic and linolic acid and erucic acid. A lower proportion of erucic acid will make the oil more palatable, nutritive besides reducing metabolic disorders.

The Protein contained in rapeseed and mustard normally ranges between 24 and 30 per cent. On the whole-seed basis and between 35 and 40 per cent on the meal basis. But the presence of toxic glucosinolates in the mustard cake is unsuitable as source of human protein and is at present use as a manual and as a cattle feed.

BACKBONE OF AGRICULTURA ECONOMY OF OILSEED CROPS

OILSEED crops have been the backbone of agricultural economy of India from time immemorial. Today these crops are cultivated on about 16.5 million hectares, with total production of 10 million tonnes. This area constitutes approximately one-tenth of the total cultivated area in India. On the oilseed map of the world, India occupies a prominent position, both in regard to acreage and production. The important oilseed crops grown in this country in order of importance are groundnut, rapeseed and mustard, sesame, linseed, safflower, castor, sunflower and niger.

GROUNDNUT

Geographical Origin

Groundnut (Arachis hypogaea L.) is believed to be the native of Brazil to

Peru, Argentina and Ghana, from where it was introduced into Jamaica, Cuba and other West Indies islands. The plant was introduced by Portuguese into Africa from where it was introduced into North America. It was introduced into India during the first half of the sixteenth century from one of the Pacific islands of China, where it was introduced earlier from either central America or South America.

Economic Importance

The oil content of the seed varies from 44 to 50 per cent, depending on the varieties and agronomic conditions. Groundnut oil is an edible oil. It finds extensive use as a cooking medium both as refined oil and Vanaspati Ghee.

It is also used in soap making, and manufacturing cosmetics and lubricants, olein steering and their salts. Kernels are also eaten raw, roasted or sweetened. They are rich in protein and vitamins A, B and some members of B2 group. Their calorific value is 349 per 100 grammes. The H.P.S. type of groundnut kernels are exported to foreign countries. The residual oilcake contains 7 to 8 per cent of N, 1.5 per cent of P 2O5 and 1.2 per cent of K2O and is used as a fertilizer. It is an important protein supplement in cattle and poultry rations. It is also consumed as confectionary product. The cake can be used for manufacturing artificial fibre. The haulms (plant stalks) are fed (green, dried or silaged) to livestock. Groundnut shell is used as fuel for manufacturing coarse boards, corksubstitutes etc. Groundnut is also of value as rotation crop. Being a legume with root nodules, it can synthesise atmospheric nitrogen and therefore improve soil fertility.

Botanical Description

Groundnut (Archis hypogaea L.) is a member of sub-family, Papilionaceae of the family Leguminosae. Archis hypogaea L. consists of two subspecies each containing two botanical varieties.

Plants of the botanical variety hypogaea are spreading (runner) to upright (erect bunch) in growth habit, have alternate branching, lack inflorescences on the main stem, possess appreciable fresh seed dormancy, flowers are longer and mature later than those of subspecies fasligiata. Variety hirsuta has been used only to a little extent.

Plants of this subspecies fasligiata are upright, have sequential branching and inflorescences in the main-stem leaf axils, possess little fresh seed dormancy, and are of shorter duration than those of the subspecies hypogaea. Subspecies fasligiata includes both the Spanish and Valencia types

Groundnut, in general, has a short-statured plant, with the main axis being upright (15 to 40 cm long) but the major part of the plant consists of the primary branches. Secondary and tertiary branches are found in the semi-spreading and spreading (Virginia) types, giving them a prostrate stature. The leaves are alternate, stipulate and quadri-foliate. The flowers are orange yellow, typically papilionaceous, with a long calyx tube within which is held the style borne on a superior ovary. The calyx and corolla lobes are borne in the axils of the leaves on the fruiting branches. They are bisexual, Zygomorphic, complete and sessile. Petals are five, with one large standard, two wings and two fused keel petals. There are two steering and eight fertile anthers, four of which are globose and four oblong (dimorphic).

Groundnut is predominantly a self-pollinated crop and pollination takes place early in the morning. As soon as the fertilization is complete, the flowers fade. After fertilization, an intercalary meristem becomes active at the base of the ovary above the point of attachment of the hypanthium, producing new tissue below itself and resulting in an elongated stalk in the peg. The pegs are positively geotropic, enter the soil and bend in a horizontal plane. Generally two, and occasionally one, three or four, fertilized ovules are borne at the tip of the peg which later swells to become the pod. The testa is generally pink, but varieties with red, white, purple and blotched testa, with various gradations of colours are available.

Distribution, Area and Production

The major groundnut-producing countries of the world are India, China, Nigeria, Senegal, Sudan, Burma and the USA. Out of the total area of 18.9 million hectares and the total production of 17.8 million tonnes in the world, these countries account for 69 per cent of the area and 70 per cent of the production. India occupies the position, both in regard to the area and the production, in the world. About 7.5 million hectares is put under it annually and the production is about 6 million tonnes. 70 per cent of the area and 75 per cent of the production are concentrated in the four states of Gujarat, Andhra Pradesh, Tamil Nadu and Karnataka. Andhra Pradesh, Tamil Nadu, Karnataka and Orissa have irrigated area forms about 6 per cent of the total groundnut area in India.

Climate and Soil

Groundnut is grown throughout the tropics and its cultivation is extended to the subtropical contries lying between 45 degrees N and 35 degrees S and up to an altitude of 1000 metres. The crop can be grown successfully in places receiving a minimum rainfall of 1,250 mm. The rainfall should be well distributed well during the flowering and pegging of the crop. The total amount required for presowing operations (preparatory cultivation) is 100 mm; for sowing, it is 150 mm and for flowering and pod development an evenly distributed rainfall of 400-500 mm is required. The groundnut crop, however, cannot stand frost, long and severe draught or water stagnation.

Groundnut is grown on wide variety of soil types. However, the crop does best on sandy loam and loamy soils and in the black soils with good drainage.

Heavy and stiff clays are unsuitable for groundnut cultivation as the pod development is hampered in these soils.

OILSEEDS SCENARIO IN INDIA

India is amongst the largest producer and consumer of vegetable oils in the World. Oilseeds have been the backbone of agricultural economy of India since long. Indian vegetable oil economy is the fourth largest in the world next to USA, China and Brazil. Oilseed crops play the second important role in the Indian agricultural economy next to food grains in terms of area and production.

The Indian climate is suitable for the cultivation of oilseed crops; therefore, large varieties of oilseeds are cultivated here. The major oilseeds cultivated in our country are Groundnut, Rapeseed and Mustard, Castor seed, Sesamum, Nigerseed, Linseed, Safflower, Sunflower and Soybean. However, Groundnut, Rapeseed/Mustard, Soybean and Sunflower account for a major chunk of the output. At present, more than 27 million hectares of land is under oilseeds cultivation. The area under oilseeds has been increasing over time and the production has registered many fold increase but its productivity is still low as compared to other oilseed producing countries in the world.

The low and fluctuating productivity is primarily because cultivation of oilseed crops is mostly done on marginal lands, which are lacking in irrigation and low levels of inputs are used here. To improve the situation of oilseeds in the country, government of India has been pursuing several development programmes. They are Oilseed Growers Cooperative Project, National Oilseed and Development Project, Technology Mission Oilseeds (TMO) and Integrated Scheme of Oilseeds, Pulses, Oil Plam and Maize.

The concerted efforts of these development programmes/schemes register significant improvement in annual growth of yield and area under oilseed crops. However, India, still imports a significant proportion of its requirement of edible oil. The Technology Mission on Oilseeds adopted a four - pronged strategy in order to harness the best of production processing and storage technologies for attaining self-reliance in vegetable oils. The mission initiated corporatization and modernization process in the oilseed sector. The efforts result the significant growth in the production of oilseed under the mission period and the highest growth was observed in the case of soybean and sunflower oilseeds. The area under total nine oilseed crops registered growth of 4.3 per cent and production registered growth of around 8 per cent, which is highest since, independence, both area expansion and technology improvement contributed almost equally to attain this growth in oilseed crops. India was about to attain self sufficiency during the mission period and import of edible oils reduced to merely 6 per cent of total agricultural imports in value terms. The expansion of irrigation facilities and transfer of new technologies help to obtain the desired results.

The total area under oilseed crops was 19.0 million hectares during 1985-86 which increased 25.2 million hectares during 1992-93 and area under irrigation also increased from 17.3 per cent in 1985-86 to 24.1 per cent in 1992-93. The total production showed remarkable jump from 10.83 million ton to 20.11 million ton during this period and import fallen drastically from 1.1 million ton to 0.1 million ton in 1992-93.

The period from mid 90s to early 2000s observed declaration in the area and output of oilseed crops. It is realized that the effects of TMO peter outs with the time and negative growth in both area and production was recorded. The area under oilseeds registered - 2.7 per cent growth and production registered - 1.8 per cent growth. It reveals that the area under oilseeds is shifted to other crops and farmers are losing interest in growing oilseed crops. The low output prices, high risk, attack of insects and pests, weather aberrations etc are some of reasons for declaration in area and output. Higher Government minimum support prices (MSP) for wheat and rice, important competring crops for oilseeds in some regions, also showed the growth in oilseed area during this period.

The oilseed crops registered momentum with the dawn of millennium. During the period of 2000-2009, the total nine oilseed crops are showing growth of 2.5 per cent, 5.2 per cent and 2.6 per cent in area, production and productivity respectively. The total oilseeds area, production and productivity was 22.7 million hectares, 18.4 million ton and 810 kg/ha respectively during 2000-01, which increased to 27.6 million hectares, 27.7 mi8llion ton and 1006 kg/ha during 2008-09. It is heartening to note that the significant growth in production is coming from the yield effect. It is welcoming sign as the area for cultivation is limited. The advance estimates reveals that the Kharif oilseed production will meet the set targets of the year 2010-11.

The significant improvement in annual growth in area and yield under total nine oilseed crops during 2000-01 to 2008-09 as compared to period of 90s has resulted in increase in the annual growth rate of production of oilseed, but at the same time the import of oilseeds oil hovering between 40-50 per cent of total agricultural imports by India. This gap is likely to increase unless more vigorous efforts to increase production and productivity of oilseeds are made. In order to reduce the import and to ensure reasonable level of self-sufficiency, the current growth rate needs to be maintained.

AGRICULTURE, FOOD AND OILSEED CROPS

About three-quarters of China's cultivated area is devoted to food crops. China is the world's largest rice producer, and rice is the country's most important crop, raised on 26 per cent of the cultivated land. Most rice is grown south of the Huai River, notably in the middle and lower Yangtze Valley, in the Zhu Jiang (Pearl River) delta, and also in Yunnan, Guizhou, and Sichuan provinces.

Wheat is China's second most important food crop. Wheat is grown in most parts of the country, but the largest growing areas are on the Huabei Pingyuan, in the valleys of the Wei and Fen rivers on the Huangtu Gaoyuan (Loess Plateau), and in Jiangsu, Hubei, and Sichuan provinces. Although the area of wheat cultivation is nearly as large as that of rice (24 per cent of cultivated land), the yield is lower.

Corn (maize) occupies 19 per cent of the country's cultivated area, mainly in northern, northeastern, and southwestern China. It is increasingly used as animal feed and less is taken for direct human consumption. Kaoliang (a sorghum) and millet are important food crops in North and Northeast China. Kaoliang is also used as an animal feed and converted into alcohol for a beverage; the stalks are use to make paper and as a roofing material. Oats are important chiefly in Inner Mongolia and in the west, notably in Tibet.

Other food crops include sweet potatoes, white potatoes, and various other fruits and vegetables. Sweet potatoes predominate in the south and white potatoes in the north. Fruit includes tropical varieties such as pineapples and bananas, grown on Hainan Island; apples and pears, grown in the northern provinces of Liaoning and Shandong; and citrus fruits, particularly oranges and tangerines, which are major products of South China.

Oilseeds play a major role in Chinese agriculture, supplying edible and industrial oils as well as other food products, and constituting an important share of exports. The most important oilseed is the soybean, which is grown mainly in North and Northeast China. Chinese soybeans are particularly good for making tofu (bean curd), and the oil made from soybeans is used in cooking. China is one of the world's leading soybean producers and is also a leading producer of peanuts, which are grown in Shandong and Hebei provinces. Other important oilseed crops are sesame seeds, sunflower seeds, and rapeseed. The seeds from the fruit of the tung tree also provide a valuable oil, which is used as an additive in paints and varnishes. More than half the tung oil produced in

China originates in Sichuan. Tea is a traditional export crop of China, and the country produces more than 20 per cent of the world supply. Green and jasmine teas are very popular among the Chinese population, whereas black tea is mostly for export. The principal tea plantations are on the hillsides of the middle Yangtze Valley and in the southeastern provinces of Fujian and Zhejiang.

China obtains sugar both from sugarcane and sugar beets. Sugarcane is grown mainly in the provinces of Guangdong and Sichuan. Sugar beets, a relatively new crop for the country, are raised in Heilongjiang Province and on irrigated land in Inner Mongolia.

10

Organic Manuring and their Role to Incise Soil Fertility Status for Higher Crop Production

ORGANIC MATTER

Soil organic matter is a very important factor in soil fertility. It is a reservoir of plant nutrients, has a high CEC, buffers soil pH, and chelates micronutrients. Organic matter exists in different forms in soil, ranging from living soil organisms to fresh, readily decomposed plant residues to humus that is very stable and resistant to further degradation.

Living soil organisms include bacteria, fungi, actinomycetes, nematodes, earthworms, mites, and insects. They make up the soil food web, which carries out biological nutrient cycling.

Plant roots are a sometimes forgotten part of the living soil biomass. Readily decomposed or active organic matter is the form of organic matter through which nutrients are actively recycled. Decomposition produces gums, polysaccharides (sugars), and other compounds that are the "glues" of water-stable soil aggregates necessary for good soil structure. Stable humus contributes to long-term nutrient supply and is the organic matter fraction with high CEC.

Chelation is the ability of soluble organic compounds to form complexes with micronutrient metals that keep them in solution and available for uptake. In organic soils (peats and mucks), trace metal complexes with organic matter can reduce their availability.

The cycling of plant nutrients through soil organic matter supplies a significant portion of a growing crop's nutrient needs. Another aspect of this cyclical process is that organic matter not only contributes to soil fertility, but fertile soils contribute to the production of organic matter. One of the best ways to add organic matter to the soil is to maintain fertility and grow healthy crops that add large amounts of plant residue.

NUTRIENT CYCLES

Soil fertility can be maintained when nutrients are efficiently recycled through the soil food web and soil-plant-animal system. Nutrient cycling is conveniently illustrated in diagrams that range from very simple to extremely complex.

BASIC PLANT NUTRIENT CYCLE

The basic plant nutrient cycle highlights the central role of soil organic matter. Cycling of many plant nutrients, especially N, P, S, and B, closely follows parts of the Carbon Cycle. Plant residues and manure from animals fed forage, grain, and other plant-derived foods are returned to the soil. This organic matter pool of carbon compounds becomes food for bacteria, fungi, and other decomposers. As organic matter is broken down to simpler compounds, plant nutrients are released in available forms for root uptake and the cycle begins again. Plant-available K, Ca, Mg, P, S, and some micronutrients are also released when soil minerals and precipitates dissolve.

NITROGEN CYCLE

The N cycle is the most complex nutrient cycle (the S cycle is equally complex). N exists in many forms, different physical states as well as both organic and inorganic compounds, so transformations between these forms make the N-cycle resemble a maze rather than a simple, circular cycle. Biochemical transformations of N, such as nitrification, denitrification, mineralization, immobilization (assimilation), and N-fixation, are performed by a variety of soil-inhabiting organisms. Physical transformations of N include several forms that are gases, which move freely between soil and atmosphere. Although the N-cycle is very complex, it is probably the most important nutrient cycle to understand.

There are two reasons for this: 1) N is usually the most growth-limiting plant nutrient in terrestrial (land) ecosystems, so there is often a very large crop-yield response to additional N, and 2) N in the nitrate form is very soluble and one of the most mobile plant nutrients in soil, so it can easily be lost from farm fields and become a contaminant in surface or groundwater. Managing N is a critical part pf soil fertility management.

BIOLOGICAL TRANSFORMATIONS OF NITROGEN

- Nitrification – conversion of ammonium-N (a cation held in soil by CEC) to nitrate-N (a soluble anion easily lost in run-off or leaching)
- Denitrification – conversion of plant-available nitrate-N to N-gases that are unavailable to plants and easily lost from soil
- Mineralization – biological breakdown of organic-N and release as plant-available ammonium-N

- Immobilization (assimilation) – uptake of inorganic-N from soil and incorporation into organic-N compounds in microbes (N becomes unavailable to plants)
- N-Fixation – conversion of N-gas in the air to organic-N that becomes available to plants (performed by bacteria associated with roots of legumes and other plants, and some free-living soil microbes)

NUTRIENT BALANCE AND NUTRIENT BUDGETS

Nutrient cycling is not 100 per cent efficient. There are always some losses or "leaks" from the cycles, even for natural ecosystems. In farming systems, where products are bought and sold, the balance between nutrient inputs and outputs is easily shifted in one direction or the other. When the balance between inputs and outputs is quantified, a nutrient budget can be calculated. Nutrient budgets can be determined at different scales, from single fields to whole farms to landscapes and even broader regional areas. Strictly speaking, a cycle is a circular, closed-loop pattern, so the nutrient cycles. There are cycles within them, but they include other components and describe a larger picture where there is movement or flows of nutrients into and out of smaller systems such as farm fields. Nutrient balances or budgets look at these nutrient flows between different systems.

WHOLE-FARM NUTRIENT BUDGETS

Different types of farms have different patterns of nutrient flow. They vary in patterns of internal movement within the farm as well as in the amounts of external transfers both onto and off of the farm. Cash crop and concentrated livestock farms represent two extremes in nutrient-flow patterns, with mixed crop and livestock farms in an intermediate position. Looking at these three farm types outlines the consequences and challenges faced by a range of different farm types in maintaining soil fertility, using plant nutrients efficiently, and eliminating uncontrolled nutrient flows off farms and into the surrounding environment.

CASH CROPS

Cash grain and vegetable farms that do not have livestock frequently export large amounts of plant nutrients in off-farm sales. A 500-cwt/acre potato crop, for example, removes about 215 lbs of N, 30 lbs of P, and 240 lbs of K in the harvested tubers. A 150-bushel/acre corn crop contains about 135 lbs of N, 25 lbs of P, and 35 lbs of K in the grain. When corn stover or small grain straw is sold in addition to grain, nutrient losses from the farm are larger, especially for K. To maintain high yields, these nutrients must be replaced. Biologically-fixed N from soybeans or other legumes in the rotation supplies some N, but large N inputs from forage legumes are not usually part of systems without livestock to consume the forage. When high quality hay is grown as a cash crop, nutrient exports off the farm are even greater than for grain or vegetables. There are some deep, naturally fertile soils with high organic matter and mineral reserves that can be "mined" and meet many crop nutrient needs for some time, but large amounts of off-farm fertilizer inputs are required in most soils for cash-crop systems to maintain nutrient sufficiency and crop yields. In this age of globalization, international grain sales have become an important market for U.S. farmers. One consequence of global trade is the associated, worldwide transfer of plant nutrients.

MIXED CROP AND LIVESTOCK

Farms with both crops and livestock have the potential to recycle a large portion of the nutrients used by crops back to the soil, because about 75 per cent or more of the NPK consumed in animal feed is excreted in manure or urine. Efficient recycling depends upon storage, handling, and application methods that minimize losses, and an effective nutrient management plan that applies manure to fields in amounts matching crop needs with the nutrient content of the manure.

Within a farm, manure applications can be a method of transferring nutrients between fields. Depending upon the balance between crop and livestock enterprises, whole-farm nutrient budgets on mixed farms include different amounts of nutrient losses in milk, meat, or eggs, and different levels of nutrient inputs from purchased feed and fertilizer.

CONCENTRATED LIVESTOCK

Concentrated animal-feeding operations import large amounts of plant nutrients in purchased grain, forage, and bedding. They are generally net nutrient importers, because purchased inputs exceed nutrient losses from milk, meat, or egg sales.

These excess nutrients accumulate in animal wastes that often create storage or disposal problems. High-density livestock operations frequently have an inadequate land base to efficiently use all the manure they generate, so there is the potential for increased risk of water contamination. As livestock operations have become larger, they have also tended to concentrate regionally, resulting in increased geographic separation between feed-grain producers and consumers.

Manures are bulky products that are difficult and costly to apply and transport long distances. In some locations it currently is not economical to recycle the nutrients in animal waste, so long-term storage rather than re-use has become the solution to the waste problem. The net result is increasing transfer of nutrients from one part of the country to another and increased dependence on purchased fertilizer inputs in grain production areas.

MAINTAINING SOIL FERTILITY

MANAGEMENT PRACTICES TO MAXIMIZE NUTRIENT CYCLING AND NUTRIENT-USE EFFICIENCY

Nutrient management can be defined as "efficient use of all nutrient sources" and the primary challenges in sustaining soil fertility are to:

- Reduce nutrient losses
- Maintain or increase nutrient storage capacity
- Promote recycling of plant nutrients
- Apply additional nutrients in appropriate amounts

In addition, cultural practices that support the development of healthy, vigorous root systems result in efficient uptake and use of available nutrients. Many cultural practices help accomplish these goals, including establishing diverse crop rotations, reducing tillage, managing and maintaining crop residue, growing cover crops, handling manure as a valuable nutrient source, composting and using all available wastes or byproducts, liming to maintain soil pH, applying supplemental fertilizers, and routine soil testing. These beneficial management practices have multiple effects on nutrient cycling and soil fertility, which make it important to integrate their use and examine their effects on the complete soil-crop system, rather than just a single component of that system. There are many good ways to farm, so different solutions or combinations of practices are appropriate for different systems to reach similar goals.

PHOSPHORUS FLOWS AND THE MINNESOTA RIVER

Phosphorus enrichment of surface waters is a major issue in some parts of Minnesota. In fresh water systems, P is usually the limiting nutrient for growth of algae and aquatic plants, so their growth is stimulated when P in run-off or eroded soil enters lakes and rivers. Algal blooms lead to accelerated eutrophication of surface waters and degradation of water quality. In extreme cases, depleted levels of dissolved oxygen in the water cause death of fish and other aquatic life.

Water quality concerns led to a recent Minnesota law restricting P fertilizer application on home lawns and other turfgrass areas. In the agricultural landscape, similar concerns are expressed about the role of agriculture in P enrichment of the Minnesota River. P loading into the Minnesota River comes from a variety of sources, including stream bank erosion, water treatment plants, and industrial activity. The extent of the contribution from agriculture is difficult to measure, but P in run-off from farm fields and P attached to eroded soil are certainly potential pathways of P delivery to the river and its tributaries.

Phosphorus flows into Minnesota are not as dramatic as those described for the poultry production areas of the Delmarva Peninsula, but there are similarities and some common pathways. Phosphate rock mined in Florida or other distant locations is processed into phosphate fertilizers that are transported to Minnesota and applied to crop fields. Some of the grain harvested from these fields becomes part of animal feeds that are shipped to places like the Delmarva Peninsula, and additional harvested products are transported for other uses, so some of the imported P flows back out of the state in exported agricultural products. However, some of the imported P accumulates in various forms and locations, and is a potential source of nutrient loading and impaired water quality in the Minnesota River if not properly managed.

Accumulation of P in manure and increasing levels of P in the soil are two ways the flow of P into the state can build up and threaten the Minnesota River or other surface water bodies if not managed efficiently. Concentrated livestock production is not as widespread as it is in the Delmarva Peninsula, but in localized areas the amount of P in manure exceeds the amount of P required to meet crop needs on surrounding cropland. Manure is a valuable resource, but fields with a long history of heavy manure application can exceed the capacity of the soil to efficiently recycle the amount of P in continual manure additions. Buildup of soil P can also occur when P fertilizer is applied at rates exceeding crop P requirements.

Efficient use of fertilizer and manure P requires sound nutrient management planning to reduce the potential for environmental problems. This includes soil testing to determine the need for P, manure analysis, proper storage and handling of manure, and fertilizer and manure application methods that reduce the potential for movement of P from farm fields. In addition, soil management practices that limit surface run-off and reduce soil erosion help protect water quality, as well as sustaining long-term soil productivity.

CROP ROTATIONS

The term 'rotation effect' was coined to describe the observation that yields for a crop grown in rotation with other crops are usually 5 to 15 per cent greater than for continuous monoculture of that same crop. The reason for increased yields is not always clear, and in most cases it is probably not due to a single cause, but growing a variety of crops in sequence has many positive effects on soil fertility. In a diverse rotation, deep-rooted crops alternate with shallower, fibrous-rooted species to bring up nutrients from deeper in the soil. This captures nutrients that might otherwise be lost from the system. Differences in plant rooting patterns, including root density and root branching at different soil depths, also results in more efficient extraction of nutrients from all soil layers when a series of different crops is grown.

Including sod-forming crops in rotation with row crops decreases soil and nutrient losses from run-off and erosion, and increases soil organic matter. Growing legumes to fix atmospheric N reduces the need for purchased fertilizer and increases the supply of N stored in organic matter for future crops.

Biologically fixed N is used most efficiently in rotations where legumes are followed by crops with high N requirements. Rotating crops also increases soil biodiversity and nutrient cycling capacity by supplying different residue types and food sources, reduces the buildup and carryover of soil-borne disease organisms and insect pests (breaks disease and pest cycles), and can help create favorable growing conditions for healthy, well-developed crop root systems.

SOIL AND WATER CONSERVATION PRACTICES

Soil erosion removes topsoil, which is the richest layer of soil in both organic matter and nutrient value. Implementing soil and water conservation measures that restrict run-off and erosion minimizes nutrient losses and sustains soil productivity. Tillage practices and crop residue cover, along with soil topography, structure, and drainage, are major factors in soil erosion. Surface residue limits erosion by reducing detachment of soil particles by wind or raindrop impact and restricting water movement across the soil. Tillage practices manage the amount of crop residue left on the soil surface. Reduced tillage or no-till maximizes residue coverage. Water moves rapidly and is more erosive on steep slopes, so reducing tillage, maintaining surface residue, growing sod crops, and planting on the contour or in contour strips are recommended conservation practices. Using diverse rotations and growing cover crops also can reduce erosion.

Soils with stable aggregates are less erodible than those with poor structure, and organic matter (including the activity of living soil organisms and fine roots) helps bind soil particles together into aggregates. Tillage breaks down soil aggregates and also increases soil aeration, which accelerates organic matter decomposition.

Well-drained soils with rapid water infiltration are less subject to erosion, because water moves rapidly into and through them and does not build up to the point where it moves across the surface. Drainage improvements on poorly drained soils reduce run-off, erosion, and soil compaction. Improving drainage also decreases N losses from denitrification, which can be substantial on waterlogged soils, by increasing aeration. Improving aeration in the plant-root zone also promotes healthy root growth. A negative consequence of improved drainage is loss of nitrate-N and other nutrients through tile outlets to surface waters. Especially important are flushes of residual N after late winter/early spring rains.

COVER CROPS AND GREEN MANURES

Growing cover crops and green manure crops can be viewed as a type of crop rotation, where adding a non-revenue generating crop between annual cash crops extends the growing season. Many of the benefits, therefore, are the same as those achieved with crop rotation.

The terms cover crop and green manure are frequently used synonymously. They perform many similar functions and many of the same plant species are used as both cover crops and green manure crops. The main difference between the two is that the primary purpose of growing a cover crop is to protect the soil surface from raindrop impact, run-off, and erosion and the primary purpose of a green manure is as a soil-building crop to produce organic material for incorporation into the soil. Winter grains like cereal rye planted after potatoes are cover crops that are designed to hold soil in place until the next main crop is planted in the spring, but they also add organic matter to the soil when they are turned under. Rapidly growing summer annuals like buckwheat and sorghum-sudangrass are planted between short-season vegetable crops as green manures to add organic matter to the soil, but they also protect the soil from erosion between spring and fall vegetables.

Growing legume cover crops adds biologically fixed N. The additional plant diversity with cover crops stimulates a greater variety of soil microorganisms, enhances carbon and nutrient cycling, and promotes root health. The soil surface is covered for a longer period of time during the year, so nutrient losses from run-off and erosion are reduced. This longer period of plant growth substantially increases the amount of plant biomass produced, which in turn increases organic matter additions to the soil. The extended growth period obtained with cover crops also extends the duration of root activity and the ability of root-exuded compounds to release insoluble soil nutrients.

A winter cover crop that makes good fall growth traps excess soluble nutrients not used by the previous crop, prevents them from leaching, and stores them for release during the next growing season. Complementary cover crop mixtures produce root exudates with varying composition and effects, and have different zones of nutrient uptake, because they differ in amount, depth, and patterns of root branching. Deep-rooting cover crops, like sorghum-sudangrass hybrids and sweet clover, can break up some types of compacted soil layers and improve rooting depth for the next crop. Cereal rye, sorghum-sudangrass, and brassicas (mustards), such as oilseed radish and forage turnip, all suppress some nematode species and may be useful cover crops in fields with moderate infestation levels. Cover crops also can suppress weeds, which otherwise would compete with crops for nutrients.

Cover crop benefits are probably greatest as soil-building crops preceding high-value perennial fruits and in rotations with low-residue, short-season crops such as annual vegetables. It is often easier find places to grow cover crops in vegetable rotations than in agronomic rotations, and there may be opportunities to grow both summer and fall cover crops in vegetable systems. Many vegetables have relatively shallow, sparse root systems, but are well fertilized because of their value. Both summer and fall cover crops absorb residual nutrients, in addition to increasing the time and amount of surface cover.

Disadvantages of growing cover crops are:

- Large amounts of residue can make planting difficult and reduce crop stands
- In wet springs, planting may be delayed if wet soil conditions delay killing the cover crop
- Soil warms more slowly in the spring under cover crops than for tilled soil and lower soil temperatures can slow seed germination, reduce early-season growth, delay maturity, and reduce crop yields
- Spring cover crop growth uses water, which can adversely affect the following cash crop in a dry year (in wet years, cover crop water use may be beneficial on poorly drained soils)
- Some cover crops attract and/or harbour pests that can damage succeeding crops. There are expenses and management time required to grow cover crops

Cover crops have many benefits, but when you grow them you need to commit time to their selection and management to fully realize their benefits and avoid potential problems. Select cover crops with characteristics that will meet your objectives and fit your rotations, and then manage them with the same attention and skill you give any other crop.

MANURE MANAGEMENT

Returning manure to crop fields recycles a large portion of the plant nutrients removed in harvested crops. On farms where livestock are fed large amounts of off-farm purchased feeds, manure applied to crop fields is a substantial source of nutrient inputs to the whole farming system. However, just as nutrients can be lost from the soil, nutrient losses from manure during storage, handling, and application are both economically wasteful and a potential environmental problem. Soluble nutrients readily leach from manure, especially when it is unprotected from rainfall during storage. N is readily lost through volatilization of ammonia, both during storage and when manure is not incorporated soon after field application. Nutrient losses from manure also occur when it is applied at rates exceeding crop nutrient requirements.

Nutrient management is the efficient use of all nutrient sources. A Nutrient Management Plan that takes all nutrient sources into account is not just environmentally sound, it is good business.

Analyze manure for its nutrient content and adjust application rates according to crop needs, soil tests, and frequency of manure applications. Avoid applying manure at rates that exceed crop requirements for any nutrient, but especially for N and P on fields that receive manure on a regular basis. This often means that rates should be based on P requirements rather than N requirements. Following heavy manure applications with crops that have high nutrient requirements (especially for N and P) reduces losses and increases nutrient-use efficiency. In addition to nutrient value, manure adds organic matter to the soil, which can improve soil structure and increase CEC.

ORGANIC CROP PRODUCTION: FERTILITY

There are 16 nutrients that all plants must have. More than 90 per cent of the dry matter weight of the plant is made up of Carbon (C), Hydrogen (H) and Oxygen (O), but these nutrients are universally available from air and water.

Of the remaining essential nutrients, plants require large amounts of Nitrogen (N), Phosphorus (P), Potassium (K) and Sulphur (S), thus they are referred to as macronutrients. A lesser amount of the secondary nutrients and very small amounts of the micronutrients are also required.

Despite differences in the amounts required by different species of plants, each of these nutrients is essential for plant growth. Small amounts of the micronutrients are just as necessary for plant growth as large amounts of the macronutrients. However, when too much of an essential nutrient (especially one of the micronutrients) is available, it may become toxic to plant growth.

Table. The essential plant nutrients macronutrients.

Supplied mostly by air and water	Supplied by the soil		
	Macronutrients	Secondary Nutrients	Micronutrients
Carbon (C)	Nitrogen (N)	Calcium (Ca)	Iron (Fe)
Hydrogen (H)	Phosphorus (P)	Magnesium (Mg)	Manganese (Mn)
Oxygen (O)	Potassium (K)		Boron (B)
	Sulphur (S)		Molybdenum (Mo)
			Copper (Cu)
			Zinc (Zn)
			Chloride (CI)

Crop yield is a function of the law of the minimum which states that yield is determined by the factor that is most limiting.

For example, if only enough N is available to grow a 30 bu./ac. crop, yet there is enough of all the other essential nutrients to grow a 40 bu./ ac. crop, the crop will still yield only 30 bu./ac. because yield will be limited by the supply of N.

The law of the minimum also includes factors other than nutrients such as moisture, light and heat. Continuing the previous example, if moisture supply was only able to produce a 20 bu./ac. crop, the crop would yield 20 rather than 30 bu./ac.

Nutrients come from two main sources, the breakdown of soil minerals and the decomposition of organic matter (mineralization). Nutrient release from organic matter breakdown begins within a few days to several years and can last for centuries. Nutrient supply from soil mineral breakdown is much slower. Depending on the mineral, it may take centuries for an appreciable amount of nutrient release. Plants take up nutrients as inorganic ions, whether the nutrients originate from soil minerals, organic matter or approved organic fertilizers. For example, N is taken up as nitrate ($N03_{-}$) and ammonium ($NH4_{+}$) ions, and P is taken up mostly as inorganic orthophosphates.

Legumes, that fix N from the air via their symbiotic association with bacteria in nodules on the plant roots, are the exception.

NUTRIENT BALANCE OR NUTRIENT RATIOS

When certain nutrients are in high concentrations in the soil, they may create imbalances in other nutrients. For example, soils with very high levels of P may induce a zinc deficiency in a flax crop. However, the use of nutrient ratios may only be useful in cases where nutrients are abundant. For example, soils low in organic matter (*e.g.*, sandy soils; eroded soils) usually have small total amounts of exchangeable cations. A deficiency may occur even though the nutrients are present in the appropriate ratios. onversely, a clay soil with a high cation capacity may not have an ideal nutrient ratio, but levels may still be adequate for crop production.

THE SASKATCHEWAN SITUATION NITROGEN (N)

Nitrogen comes from the breakdown of organic matter. In Saskatchewan, soils typically have 15 to 50 lb./ac. of plant available inorganic N released per year through the breakdown of soil organic matter and crop residue. A 30 bu./ ac. wheat crop with about 13 per cent protein content requires about 57 to 70 lb./ac. of N. Most N in organic production results from growing properly inoculated legumes as cash crops or green manures. The amount of N supplied depends on the type of plant and the plant materials incorporated into the soil. Animal manures and compost are both good sources of N available for use on organic farms. Organic producers are cautioned to check that all inputs, input ingredients and processing methods meet certification standards.

PHOSPHORUS (P)

Saskatchewan soils contain large amounts of total P, often 400 to 2,000 lb./ac. in the zero to six inch depth. About half of the total P is present as organic compounds and about half as inorganic minerals.

Unfortunately, most of this P is unavailable to plants due to the relatively high pH levels in our soils and the resultant low solubility of P minerals. In fact, Saskatchewan soils are inherently low in plant available P. A survey of organic fields revealed that most had deficient levels of plant available P. As plants absorb P from the soil solution, it is replaced by the dissolution of inorganic P minerals and the breakdown of organic P compounds. However, this process is often too slow to meet the needs of a growing crop.

Furthermore, how long soil minerals and organic matter can effectively maintain enough P in the soil solution for plant growth is dependent on a number of factors.. If the soil is no longer able to supply enough P, fast enough to meet the needs of the growing crop, yields will decline unless plant available P is added.

Organically managed soils tend to be low in plant available P because levels of replacement are usually less than that removed with harvest. However, there are several sources of P for use on organic farms, including animal manures and rock phosphate (RP).

Microbial inoculants and some green manures may also enhance P availability.

POTASSIUM (K)

The amount of potassium in Saskatchewan soils ranges from 100 to 50,000 lb./ac. depending on the parent material from which the soil has been formed. However, usually less than two per cent of the total K is plant available.

It is estimated that in Saskatchewan up to one million acres are K deficient, even though potash (K plus Cl) is mined below ground. The most severe deficiencies occur in the north and northeast, but K deficiency is possible in localized areas in all regions of the province. Sandy soils and high organic soils are the most likely to be K deficient.

Depending on the type of plant, the amount of K required can be as high as or even higher than the amount of N required. Therefore, although K nutrition may not presently be of concern over most of the province, it is very important for those soils that are or will soon be deficient. Organic producers are cautioned to check that all inputs, input ingredients and processing methods meet certification standards.

SULPHUR (S)

Although S deficiencies occur most often in the Grey and Black soil zones, S deficiencies for high S using crops have been reported in other soil zones in Saskatchewan. Sulphur, like P and other nutrients, comes from decomposition of organic matter and weathering of soil minerals.

For the first half-century or more of cultivation, S deficiency was not a concern on most of our soils. This is because, even on soils with a low S parent material, a huge pool of organic S was available as the organic matter from decaying prairie grasses was mineralized.

However, decades of cropping, particularly with high S-using crops such as canola, mustard and alfalfa, has left some Saskatchewan soils S deficient, particularly the sandy Grey and Black soils in the northern grain belt.

Animal manure generally contains low levels of S as compared to other nutrients and can serve as a small source of S. A manure analysis available from a soil test laboratory can provide amount and balance of nutrients for a manure sample. Organic producers are cautioned to check that all inputs, input ingredients and processing methods meet certification standards.

OTHER NUTRIENTS

At this time, deficiencies of the other essential nutrients are not widespread in Saskatchewan. However, a number of localized deficiencies have been identified and more may show up with time, so soil testing for all nutrients every few years is recommended. For example, copper (Cu) may be deficient in sandy or peaty Grey soils.

NUTRIENT LEVEL DECLINE

Saskatchewan soils have lost approximately 41 to 53 per cent of the organic matter; 31 to 56 per cent of the organic N; and 40 to 67 per cent of the potentially mineralizable N, depending on the soil, since cultivation began. Saskatchewan currently exports almost all the grain grown. Thus, native soil nutrient levels will continue to decline over time and nutrients need to be added to maintain yields. One of the best ways to provide adequate levels of nutrients is to build up soil organic matter content.

SOIL TESTING

Soil testing is an important tool in farm management and should be performed annually. It provides the producer with the approximate level of nutrients in the soil.

Soil testing becomes particularly important when soil nutrient levels are deficient or marginal and informed decisions need to be made. Soil testing over a number of years can also allow the producer to measure the benefit of various agronomic practices. Each soil-test laboratory has specific sampling procedures and will provide them upon request.

Nutrient levels can vary considerably across a field, particularly in rolling terrain. Knolls are often the least fertile, whereas low areas are often the most fertile.

Therefore, when taking soil samples, it is important to take them from average areas of the field to obtain an average representation of the field nutrient level. Some producers soil test the knolls, mid-slopes and low areas of a field separately for site specific management. For example, manure can be applied to knoll areas first, then to mid-slope areas and may not be needed in the low slope areas.

SUPPLYING NUTRIENTS ORGANICALLY

How, when and which products to use can be difficult decisions. Providing organic nutrients successfully requires a high level of management.

The products and practices available to organic producers vary somewhat depending on the organic certification association to which they belong. Producers reading this bulletin should contact their certification agency to ensure that products and practices are acceptable.

The main products and practices that will supply crop nutrients organically are the following:

- Applying animal manure;
- Green manuring or greenfallowing;
- Growing properly inoculated legumes in the rotation;
- Rotating high and low nutrient demand crops;
- Rotating deep and shallow rooted crops;
- Applying acceptable organic crop nutrient products;
- Use of registered nutrient solubilizing microbial inoculants, such as JumpStart®;and
- Summerfallowing using conservation tillage.

APPLYING ANIMAL MANURE

Applying animal manure to a field is an effective way of increasing both nutrients and organic matter content in the soil. Animals typically pass, as waste, 75-90 per cent of the nutrients they are fed.

The level of the various nutrients in the manure varies according to the type of animal, type of feed, how the manure is stored and how it is applied. However, animal manure contains some level of all the essential nutrients.

Soil testing laboratories can test manure for its nutrient content and the soil from the field where the manure will be applied so the correct amount can be applied to correct the nutrient deficiencies.

The organic certification agency may require that manure be composted before application. Properly composted materials reach internal temperatures high enough (ideal temperature range is 54 to 60 C) to kill pathogenic bacteria and weed seeds.

However, temperatures should not exceed 66 C or the microbes responsible for the composting process will also be killed. Overheating is prevented by turning the pile. Composting stabilizes the N in raw manure, thereby reducing N losses. Nutrients from compost are released slowly and steadily for plant use throughout the season. Composting also reduces the bulk of the material reducing transportation costs.

Another benefit of composting animal manure is that the compost process helps solubilize rock phosphate (RP). The composting process produces organic acids. When RP is mixed into manure compost, the organic acids may solubilize some of the RP, producing more plant available P.

This process can be further enhanced by adding elemental S to the manure compost. Micro-organisms in the compost may convert some of the elemental S to plant available sulphate-S. Also, during the conversion of elemental S to sulphate-S, acid is produced which may also solubilize some of the RP, producing more plant available P.

GREEN MANURING OR GREENFALLOWING WITH ANNUAL LEGUME CROPS

Green manuring involves using crops as sources of nutrients. Greenfallowing, a new term to describe a specific green manuring practice, uses properly inoculated annual legume crops such as: pea, lentil and chickling vetch for green manure during summerfallow. Green manure improves soil health by increasing soil organic matter, which in turn increases soil aggregation, as well as the nutrient supplying and water-holding capacity of the soil.

Properly inoculated legumes such as sweetclover, Indianhead lentil and chickling vetch are frequently grown to supply N to the soil. Green manures are usually grown in place of summerfallowing a field, protecting the soil from wind and water erosion. However, they must be carefully managed and incorporated early enough that they don't unnecessarily deplete the soil moisture for subsequent crops, but late enough that they provide optimum amounts of organic matter and N to the soil. Other crops (*e.g.*, fall rye) are grown to assist with weed and disease control. During greenfallow the growth stage that the annual legume is terminated is usually determined by the amount of available moisture: if moisture is limiting the annual legume is terminated early. If moisture is adequate, the crop is terminated at about the full bloom stage to optimize biomass and N fixation.

Sweetclover is a biennial that is inoculated and usually underseeded in a grain crop one year and incorporated at the bud to 10 per cent flower stage the following year. In its second year of growth, sweetclover is a high moisture using crop. Since most of the N fixation has occurred by the bud stage, green manuring sweetclover at the bud stage will maximize N accumulation and allow as much time as possible for recovery of moisture reserves and for residue decomposition. Sweetclover may not be an option when moisture is limiting, for example, dry areas of the Brown soil zone.

Inoculated Indianhead lentil is normally sown (at approximately 35 lb./ac. of seed) soon after the risk of severe frost has passed (*i.e.*, just before wheat is sown). Indianhead lentil is a poor competitor with weeds. Early seeding allows the crop to get a jump on weed growth and allows time for replenishment of soil moisture reserves after it is incorporated. Indianhead lentil is green manured at full bloom, by which time most of the N fixation has occurred. In dry years, Indianhead lentil may not produce enough dry matter to supply significant organic matter and nutrients to the soil. Inoculated chickling vetch has a high N-fixation rate and vigourous growth, making it ideal as a green manure crop. Inoculated chickling vetch can be seeded early at about 70 lb./ac. As it performs reasonably well under moisture stress conditions, chickling vetch can be grown in drier areas of the province. Chickling vetch should be incorporated at full bloom.

The main nutrient benefit of green manuring legumes is to increase soil N levels. For example, sweetclover grown at White Fox, Saskatchewan and green manured at the early bud stage, contained 68 lb./ac. of N. This N will be made available over a number of years as microbes break down the sweetclover residue. However, legume green manure is high in N, thus the N cycles to a plant available form much faster than does the straw of cereal crops.

Legumes must be properly inoculated with the correct bacterial strain for N fixation to occur. Some organic producers grow mycorrhizal crops for green manuring. Microbes referred to as arbuscular mycorrhizal fungi colonize the roots and produce tiny structures in the roots that allow for bi-directional nutrient exchange. The plant feeds the fungus and the fungus provides the plants with nutrients, such as P, and helps it to resist diseases better and overcome stress. Thus, increased amounts of plant nutrients, such as available P, can be brought into the organic pool by growing and green manuring these crops. Management of arbuscular mycorrhizal populations is a very important tool in nutrient cycling. Lentil, pea, bean, wheat, barley, corn, sunflower and flax are examples of mycorrhizal crops. Reducing tillage, especially during summerfallow, helps minimize the reduction of the arbuscular mycorrhizal population.

Exploiting the different rooting depths of crops can also be useful in supplying nutrients to subsequent crops. For example, deeply rooted plants like alfalfa can extract nutrients from soil depths not accessed by the shallower roots of some annual crops. Although this provides a short-term solution to nutrient supplies, in the long-term nutrients will become depleted deeper in the soil. Because grain or forage is harvested and exported, nutrients other than N (that can be fixed from the air) will eventually have to be replaced from external sources such as manure and organically approved products. Organic producers are cautioned to check that all inputs, input ingredients and processing methods meet certification standards.

GROWING LEGUMES IN THE ROTATION

The main benefit of growing annual legumes, such as pea and lentil, in a rotation is that if they are properly inoculated with rhizobia they will fix 50 to 90 per cent of the N they require from the air.

Table. Nitrogen fixation in inoculated legumes grown under irrigation in southern Alberta

Plant-N derived from the atmosphere Legume		
	(%)	(lb./ac.)
Alfalfa	80	267
Sweetclover	90	223
Faba bean	90	267
Field pea	80	178
Lentil	80	134
Soybean	50	134
Chickpea	70	108
Field bean	50	62

From: Rennie, Agriculture and Agri-Food Canada Research Station, Lethbridge.

Table. Nitrogen fixation by the difference of (the inoculated plant N minus un-inoculated plant N) for four legume green manure crops at full bloom at Swift Current, SK.

Legumes	% plant N derived from atmosphere	Total fixed N (Tops and Roots) Mean lb./ac.	Total fixed N (Tops and Roots) Range lb./ac.
Chickling vetch (variety ACGreefix)	74	53	30-83
Black lentil (variety Indianhead)	65	31	19-37
Feed pea (variety Sirus)	68	40	13-60

Table. Increase in the available N of Dark Brown soils due to the residue of one annual legume crop

Legume	Available N increase(lb./ac.)
Lentil	9
Field pea	22
Faba bean	36

Legume residue has a higher level of N, and breaks down much faster than non-legume residue, resulting in faster cycling of N to subsequent crops.

From: Slinkard, Crop Development Centre, Saskatoon

Perennial legumes, such as alfalfa, can supply substantial amounts of N to subsequent crops. Alfalfa produces about as much root material as it does top growth, and a large part of the root system renews each year. Thus, a substantial amount of high N organic matter remains in the soil even though much of the top growth may have been removed from the field. A three to five year old hayed alfalfa stand can provide 50 to 100 lb./ac. (or more) N (in total, not per year) to subsequent crops over the next few years.

High soil N levels can inhibit N fixation by legumes, because legumes will preferentially use soil N before they fix atmospheric N. Thus, organic producers can make the most efficient use of resources by growing non-legume crops on animal manured, green manured or summerfallowed fields and growing legumes on fields with low levels of N, thereby allowing the N-fixing benefit of the legumes to be realized.

Phosphorus nutrition is very important for high N fixation in legumes.

ROTATING HIGH AND LOW NUTRIENT DEMAND CROPS

Different crops require different amounts of the various essential nutrients. Rotating high and low nutrient demand crops may avoid placing too high of a demand on one or more of these nutrients. For example, if soil S is low, mustard should not be grown too soon after alfalfa (unless S products have been added) since both are high S users. Alternatively, if P and K are low and either pea or faba bean can be grown, pea should be a better choice because it requires only half the P and K for the same yield. The nutrient requirements of crops, refer to Nutrient Uptake and Removal by Field Crops in Western Canada fact sheet.

APPLYING ORGANICALLY APPROVED NUTRIENT PRODUCTS

Once again it must be emphasized that producers check with their organic certification agency regarding the specific nutrient products that are allowed.

Rock phosphate (RP) use has grown in organic production. Most of the P in RP is chemically similar to the existing slowly available forms of parent material P found in our soils. Thus, the amount of available P in RP is generally low, except in acidic soils (pH less than 7). Most Saskatchewan soils have high levels of Ca and have a neutral or higher pH, so high rates of RP must be applied to meet crop demand.

Although the addition of RP will not significantly improve crop growth (except at high application rates) on most Saskatchewan soils, note that:

- the more finely ground the RP, the more plant available the P;
- P availability depends on the origin (chemical make-up) of the RP;
- P from RP is more available on low pH soils; and
- a combination of RP and a P solubilizing microbial inoculant may enhance the P availability.

RP sources should be compared on the basis of the per cent of available-phosphate, not total-phosphate. Soil testing laboratories can test RP for per cent available phosphate.

ORGANIC CROP PRODUCTION REQUIREMENTS

When considering organic production as a farming technique one must first come to realize and accept the reality of crop loss due to uncontrollable pest problems. There are times when non-chemical controls are just unacceptable. To help understand and accept these facts, it is advisable that a potential new producer to visit and tour an existing organic operation, attend organic grower meetings, and, thoroughly research the topic prior to planting the first seed. Additionally, a new producer must realize that they will be a learning curve in which 3 – 5 years maybe needed to adequately develop an understanding of the methods required to effectively produce safe foods naturally that will meet the market quality demands.

SOIL MANAGEMENT

Maintenance of good soil health is the foundation of organic production. One of the most highly touted means of achieving good soil health is through increasing soil organic matter. Since most Texas soils contain 1 per cent or less organic matter, this can be a long term, highly challenging process. High soil temperatures can cause rapid decomposition of applied organic matter, requiring large volume applications over long periods of time. Studies have shown that it is unreasonable for a grower to expect to increase soil organic matter by more than 1 per cent.

For instance, an acre of dry soil six inches deep weights approximately 2,000,000 pounds. To increase the soil organic matter content by 1 per cent will require the application of 20,038 lbs/A (460 lbs/1000 2 feet). Unfortunately most organic matter sources contain 25 per cent or more ash (inorganic matter). Consequently, to achieve a 1 per cent increase of actual organic matter with a source comprised of 50 per cent moisture and 25 per cent ash, 80,000 lbs/A would need to be applied and incorporated into the soil profile. To further challenge a producer, this will be needed on an annual basis! It is understandable how such volume application can pose real delivery, distribution and handling problems. Fortunately, benefits from applied organic matter can be achieved at much lower rates. As little as 8712 lbs/A applied organic matter has been shown to improve soil tilth, and, 21,780 lbs/A, improved plant growth.

Much of the response to applied organic matter is in its ability to improve; water retention and infiltration rate, soil aggregate stability, cation-exchange capacity, soil biological activity, as well as, serve as a reservoir for organic acids and plant nutrients such as nitrogen. The end result is a more favorable soil microclimate for plant root development and function. This in turn results in improved plant growth and yield. However, large quantities of organic matter may stimulate soil borne pathogens and insects such as wireworm, cabbage maggot, and whitefly that can cause serious damage to crops.

NUTRIENT MANAGEMENT (SOIL FERTILITY)

Maintenance of good soil health and optimal growth is depended upon proper soil nutrient management. Therefore, the major objective of a nutritional management programme is to fertilize soils in such a manner to have a sufficient quantity of nutrients in the soil available to plants so that any one does not become yield limiting. The use of soil tests is suggested as a means of determining nutrient availability status in soils. The results from these tests will provide a producer with guidelines on how much fertilizer material will be needed for successful crop growth and yield. It should be remembered that soil tests are site specific and any given soil sample should not be used to base fertilizer needs for soils in additional fields or farms not tested.

The primary element of concern in vegetable production is nitrogen. Not only is this element normally required in large quantities to satisfy most crop needs, it is highly mobile in the soil and is easily lost due to leeching. As a result, nitrogen levels can be undergoing almost constant changes in the soil. This is especially true for light, sandy soils and in high rainfall areas.

Unfortunately, soil tests generally are not reliable indicators for nitrogen availability or nitrogen fertilizer requirements. With organic production where nitrogen fertilization or availability is depended upon decomposition of organic matter and/or fixation by living organisms, soil test for this element may be even less reliable then with conventional production. However, through experience gained from supplemental fertilization and crop response, soil testing can be a helpful tool for nitrogen need determination. A complement to soil analysis for monitoring plant nutrient status during the growing season is tissue analysis. Tissue for analysis can be taken any time during the growing season. If results show deficiencies of a given nutrient is present in the plant(s), a grower can then take steps to correct the situation by adjusting his fertility programme.

FACTORS IMPACTING SOIL NUTRIENT AVAILABILITY AND CROP PERFORMANCE

There are many factors and conditions that impact plant nutrient needs. The more important factors and how they impact plant response.

Table. Guide to Interpreting Soil Test Results

Element	Crop	Element ratings (ppm)				
		V. Low	Low	Moderate	High	V. High
Nitrogen	All	0-5	6-25	26-50	51-80	
Phosphorus	All	0-5	6-10	11-20	21-40	>40
Potassium	All	0-69	70-299	30-419	420+	>300
Calcium	All	0-179	*180-459	460-749	750-3560	>3560
Sodium	All	0-98	99-399	400-999	*1000-3000	>3000
Magnesium	All	0-49	*50-99	100-150	>150	——
Sulfur	All	0-7.99	8.00-15.99	16.00-24.99	*>25.00	——
Manganese	All	——-	0-0.10	*1.00-1.49	>1.50	——
Copper	All	——-	0-0.10	0.11-0.15	>0.16	——
Boron	All	——-	0-0.39	*0.40-0.59	>0.60	——
Iron	All	——-	0-3.19	3.20-4.19	*>4.20	——
Zinc	Corn	——-	0-0.30	0.31-0.80	*>0.81	——
		None	Slight	Moderate	High	V. High
Salinity	All	0-600	601-1200	1201-2000	2001-3000	>3000

Note:

* Indicates the soil test level at which addition of a nutrient is suggested or a comment statement is generated.

Reference: Soil, Plant and Water Testing Laboratory-Texas AgriLife Extension Service. Predicted crop response to fertilizer where soil test indicates element to be: Very high = no crop response; High = no crop response; Medium = 75-100 per cent of maximum expected yield predicted without fertilization; Low = 50-75 per cent of maximum expected yield predicted without fertilization; Very low = 25-50 per cent of maximum expected yield predicted without fertilization;

Table. Factors impacting nutrient needs of plants.

Factors	Importance
Aeration	Roots require oxygen for respiration and nutrient uptake
Light	Nutrient uptake is an indirect result of photosynthesis, cannot occur with light
Moisture	Needed to dissolve nutrients in order to be absorbed by roots
Temperature	Influences growth, which in turn influences rate of uptake, more growth greater the need.
Plant age	Older the plant the less the absorption efficiency
Root system	The more extensive the root system the greater the potential for nutrient absorption
Organic matter	The reservoir for nutrients, buffer for leeching
Microorganism	Breakdown organic matter to release nutrients
Soil content	The greater the inherent content the less the supplemental needs
Soil type	The heavier the soil the greater the ability to store nutrients
Soil pH	Regulates soil supply availability
Pest and Disease	Can disrupt organs such as roots and lessen the absorption potential

The development of a sound nutrient management programme should be based on a knowledge of crop needs and soil nutrient removal ability, cation ion exchange capacity of a given soil, and the role of soil pH on nutrient availability and plant response.

Although each crop has its own specific requirement, all horticultural crops utilize N, P and K in a similar ratio. As such, supplemental nutrition should be made available in this ratio if possible.

Table. Average N – P – K content of horticultural crops.

Crop Type	% N	% P	% K	N:P	N:K
Ornamental	2.0 – 6.0	0.2 – 0.7	1.5 – 3.5	10.1:1	1.6:1
Fruit	2.0 – 7.2	0.15 – 0.3	1.0 – 2.5	10.1:1	1.5:1
Vegetable	2.4 – 5.6	0.3 – 0.7	1.5 – 4.0	10.1:1	1.5:1

The ratio of these elements found in tissue samples across plant types (fruit, vegetables, and ornamentals). Based on this data, N, P, K is found in all horticultural plant species in a 5-1-2 ratio. Where possible, soil nutrient availability should be in this ratio as well. The data for soil removal by a given yield of vegetable crops follows this ratio to a certain extent.

FERTILIZER RATIOS NEEDED TO EQUAL LEAF ANALYSIS: 15 – 3.5 – 11.25

Cation exchange capacity (CEC) determines availability of nutrients in soils. Cations are positively charged particles(ions) of elements, the most important of which are Ca, Mg, Fe, NH4, Na and H2. Clay soil particles and organic matter have negatively charged ions. As a result, the cations of the various plant nutrients can be attached to and held on the surface of these soil particles. CEC is a measure of the quantity of cations that can be absorbed or held by a soil. Soil organic matter has a very high CEC. Thus, soils with high organic matter typically have a higher CEC than soils with low organic matter. These soils also typically can store and supply nutrients to plants more readily than soils with a low organic matter content.

Table. Typical CEC of some soil texture classes.

Soil Texture	Typical CEC range (meg/160 g)
Sand	2 – 6
Sandy loam	3 – 8
Loam	7 – 15
Silt loam	10 – 18
Clay and Clay loam	15 – 30

Soil nutrition availability is also influenced by soil pH. Simply stated, pH is a measure of soil acidity or alkalinity. Technically, pH refers to the hydrogen ion concentration within a soil. The pH scale ranges from 0-14. A value of 7 represents a nutireal soil, values above 7 represents an alkaline or basic soil, and, values below 7, an acid soil. Contrary to popular belief, the pH scale is not linear but logarithmic in scope. Consequently, a soil with a pH of 6 is 10 times more acid than on with a pH of 7 and one with a pH of 5 is 100 times as acid as one with a pH of 7. In itself pH is almost non-consequential because it has no direct effect on plant growth. However, pH can impact plant growth based on its influence on the availability of essential plant nutrients and on the concentration of elements toxic to plants. Optimal pH for most vegetables range between 6 and 7.

In some situations soil pH may require adjustment. To lower soil pH, sulfur applications are recommended, whereas, to increase pH, applications of lime or limestone are suggested. Quantities of sulfur and limestone needed to adjust pH. In many areas of Texas, high pH is the problem. Unfortunately, the water sources in most of these areas are also prone to be high in free calcium carbonate.

Table. Nutrient deficiency symptoms induced by pH problems.

Element	Deficiency Symptoms
Phosphorous	Stunting, slow growth, delayed maturity, short internodes, purple or dark green foliage; old leaves die back. More severe in cold weather
Potassium	Young plants have dark green leaves with small stems and short internodes. Older leaves scorched on margins; weak stem; fruit shriveled, uneven ripening. Young leaves can crinkle and curl
Boron	Tip of growing plant turns inward and dies; bud becomes light green; roots are brown in center; flowers do not form. Leaves are small, crinkled, deformed with areas of irregular areas of discoloration
Calcium	Young leaves turn yellow then brown; growing tip bends; weak stem; short dark roots.
Iron	Young leaves are yellow between veins on older leaves. Initiates first from top to bottom; veins, margins, and tips stay green.
Magnesium	Leaves are thin, lose green color from between veins on older leaves. Start out from bottom of plant up; tend to curve upward.
Manganese	Tissue between veins turns white; leaves have dead spots; plant is dwarfed. Starts out on very young leaves
Zinc	Terminal leaves are small; bud formation is poor; leaves have dead areas. Yellow intervein mottled regions.

Therefore, the addition of sulfur to lower pH has little or no effect on reducing pH because large quantities of Ca are often applied with subsequent irrigation. As a result, the sulfur is neutralized. Producing crops with a high tolerance of high pH is the only solution to the problem.

Table. Correcting Soil pH -Amount of Lime Required to Increase pH.

(ECC) Effective Calcium Carbonate Equivalent by soil test	Lime (ECC) (lbs./A)
5,000+	4,792
3,000	3,920
2,000	3,049
1,000	1,089
500	305

Note:

**Neutralizing equivalent based on calcium being 100 per cent.*

Table. Amount of Sulphur (95 per cent S) Needed to Lower the Soil pH

Broadcast application to whole soil mass (6-inch depth)			
Soil pH found by measurement	**Sandy Soils**	**Loamy Soils**	**Clayey Soils**
7.5	435-655	655-872	872-1089
8.0	1089-1307	1307-1743	1743-2178
8.5	1743-2178	2178-2614	2614-3267
9.0	2178-3267	——	——

FERTILIZING CROPS ORGANICALLY

Once a crop's nutrient needs are determined, the next decision to be made is designing an effective fertility programme is what source for nutrients best fit the overall production system. The nutrient sources most often used in organic production are: green manure and cover crops; manure; compost; and sludge. Availability and cost usually determines which of the above is used on a given crop. However, a sound fertility programme will include green manure and cover crops in combination with one or more of the other sources. Green manure and cover crops containing a N-fixing legume are the most economical and beneficial means of supplying nutrient. These crops are usually planted in rotation with economic or cash crops and serve to improve soil tilth and water holding capacity as well as to replenish soil nitrogen and other nutrients.

A green manure crop is so called because it is one, which is planted for the purpose of plowing into the soil while still green and prior to harvest maturity. Depending upon species, some green manure crops may require as many as 120 – 160 days of growth in order to fix 100 – 200 lbs N/A. Research suggest that available N from a green manure crop will increase over a 4-6 week period following soil incorporation and then return to pre-incorporation levels. Therefore, crops following a green manure crop may need to receive supplemental applications of N from another organic source such as compost or manure tea. Legumes are the green manure crops of choice due to their efficiency in fixing N. These crops can add as much as 30 –125 lbs N/A if properly grown and managed.

A cover crop is one which is usually planted during the cropping season in which a field is usually left fallow. Such crops may or may not be harvested as a cash crop. Although non- legume cover crops do not contribute much N, they can trap it and mine other nutrients from deep with in the soil profile and bring them to the surface where they can then become available to the following cash crop once they are plowed into the soil. Some research also suggests that cover crops maybe key to the development of humic acid fractions of the soil. Increasing humus in soils is one of the primary precepts of organic production. Decomposing organic matter can also contribute P, K, S, Ca, and Mg to the soil nutrient pool. An additional benefit of green manure and/or cover crops is that they tend to reduce N leeching.

Table. Green manure crops for use in enhancing soil health.

Crop	Season	Seeding rate (lbs/A)	Type	Nitrogen value (lbs/ton dry matter)
Buckwheat	Summer	75	Non	14
Crimson	Winter	15	legume	45
clover	Winter	75	Legume	21
Rye	Summer	90	Non	60
Cowpea	Summer	75	legume	46
Sorghum	Summer	25	Legume	28
Sudan	Winter	30-50	Legume	62
grass	Winter	75	Non	20
Vetch			legume	
Wheat			Legume	
			Non	
			legume	

The use of cover crops, however, can have some adverse effects in cropping systems. They can serve to deplete soil moisture supplies, limit options for cropping sequences, and temporarily immobilize plant nutrients, increase pest problems and increase production cost.

Therefore, the key to effective and profitable use of such crops lie in creative management designs to enable a producer to take advantage of their benefits within a rotation without missing income opportunities as a result of a missed cash cropping season. Animal waste or manure is the oldest fertilizer source used by man to produce a crop.

These waste products are still widely used today and are the backbone of organic fertilization programmes. The more commonly used manure and their relative nutrient content.

Although manure is a good organic fertilizer it should be remembered that the use of fresh manure is to be avoided. Ammonia is released during the decomposition of fresh manure, which can be injurious to plants. In addition, fresh manure contains a large quantity of moisture which cause problems in handling and uniformity of distribution in a field. It also tends to increase the cost of transportation and handling. Aged manure is a better choice but problems with uniformity of distribution still occur. Therefore, properly composted manure is the most desirable choice for use in crop production. The composting process, if properly employed, reduces the moisture content and kills most of the harmful bacteria. Pulverization of the composted organic fraction reduces transportation costs and improves uniformity of distribution. Although compost is a relatively economical source of plant nutrients, composts can be quite variable in nutrient content depending upon the source and materials used in the process. This situation can prove to be a challenge to a producer in determining its composition and to determine how to use it effectively. Properly composted animal waste is essential to prevent human health problems. Research has confirmed that the human pathogen, Esehercichia coli 0157:W7 contaminated soil from contaminated manure can be transmitted to produce grown on this soil.

There are food safety issues arising from the use of manure in that animal feces contain high levels of human pathogenic organisms which can be transferred to crops on which it is used. Aged or properly composed manure tends to reduce the risk from the use of animal waste materials. However, it is important that all farms using manure follow good agricultural practices to reduce any microbial risk that may exist. These include:

- Consider the source, storage, and type of manure
- Store manure as far away as practical from areas where fresh produce is being grown and handled. If manure is not composted, age the manure at least six months prior to field application.
- Where possible, erect physical barriers or wind barriers to prevent run-off and wind drift of manure particles.
- Store manure slurry for at least 60 days in the summer and 90 days in the winter before applying to fields.
- Compost manure using proper temperature and turning techniques
- Plan manure application in a timely and careful manner
- Apply manure in the fall or at the end of the season to all planned vegetable fields preferably when soils are warm, non-saturated, and cover cropped
- Use only properly decomposed manure on crops like lettuce and leafy greens
- Avoid planting root or leafy crops in the year that manure is applied to a field
- Incorporate manure into the soil
- Do not harvest vegetables until 120 days after manure if possible.
- Document rates, dates and locations of manure applications

In addition to animal waste, commercial composting operations often utilize other waste products such as plant debris, tree and shrub trimmings, discarded food, and food processing waste. The actual elemental content of these products will determine the nutritive value of the resulting compost. As a result, consistency in nutritive value often becomes a problem. Poor quality or immature compost can tie up N in the soil and decrease N availability to plants. C:N ratio of a compost is an important consideration. During composting microorganisms require carbon for growth and energy for protein synthesis. Decomposition of organic matter wastes depends on proper balance of C and N. Rapid decomposition occurs when C:N ratio is between 15 and 35:1. Favorable ratio results in the loss of NH4(ammonia) while higher ratios can slow the process. Other factors affecting proper composting are; temperature, pH, and oxygen supply, Optimum composting conditions are: moisture content between 40 – 60 per cent; temperatures between 55 – 600 C; pH 5.0 – 9.0, and, 30 per cent free air content. Compost quality factors include age, moisture content, particle size, pH, salt concentration and purity (volume of sand, soil and other non-organic materials).

A misnomer surrounding the use of organic verses inorganic fertilizer is that the organic sources are better for plant growth and the environment. Claims are often made that plant responses are more dramatic with the use of organic sources. In actuality, the source of N etc is irrelevant to a plant. For instance, N has to be in a certain form, nitrate (NO_3) or ammonium (NH_4), before it can be taken up by plants. Consequently, if an organic source is used the N has to be converted to the nitrate form just as it has to be when inorganic forms are used.

Table. Elements and chemical forms available to green plants

Element	Form available to green plant
N (nitrogen)	N03- (nitrate ion), NH4 + (ammonium ion)
P (phosphorous)	HPO4 +and H2PO4- (mono and dihydrogen phosphate ions)
K (potassium)	K+ (potassium ion)
Ca (Calcium)	Ca++ (calcium ion)
Mg (magnesium)	Mg ++ (magnesium ion)
S (Sulfur)	S04 +and S03 +(sulfate and sulfite ions)
Fe (iron)	Fe ++, Fe +++ (ferrous and ferric ions)
Mn (manganese)	Mn ++, Mn +++ (manganese ions)
Cu (copper)	Cu +, Cu ++ (cuprous ions)
Zn (zinc)	Zn ++ (zinc ion)
B (boron)	B03 C (borate ion)
Mo (molybdenum)	Mo04 +(molybdenum ion)
Cl (Chlorine)	Cl – (choride ion)
Co (cobalt)	Co ++ (cobalt ion)

The essential elements for plant growth and their forms available to green plants. Noticed improvement in plant response with inorganic forms is really due to the effect of organic matter on soil tilth, aeration and water holding capacity of the soil and not to N. Once in the plant the source of N has no baring on the use of this element in growth. In some instances, natural or organic sources can actually contain dangerous levels of contaminants such as salts, boron and heavy metals. Such contaminants can have a disastrous effect on plant growth and yield. With regard to environmental friendliness of organic or inorganic sources, organic sources have just as high a potential to cause N leaching into the ground water as does inorganic sources. What determines the rate of leeching is the quantity and use patterns of the fertilizers.

Often times during the production of a crop, additional N is required during the growing season. Supplemental N can be applied in a side or top dressing with a good quality compost. Another method of supplying supplemental N during the growing season is through the application of manure tea. The use of this product may or may not have merit in large scale farming operations. Adding an organic source such as chicken manure or steamed bone meal to water, stirring the mixture for several days, and then draining off the liquid makes manure tea. The tea is then applied to crop either as a soil drench, band or injected through a drip irrigation system. The basic formula for the above is three pounds of manure/25 gallons of water.

Sewage sludge has also been used as a organic fertilizer. Sludge is the solid material removed from sewage treatment plants. It generally is available in three forms; raw, digested, and, activated (previously treated sludge that has been aerated so that aerobic decomposition can occur). Of these only the activated sludge should be considered for use in limited situations in agricultural crop production. The risk of contaminating food with human pathogen is a major concern with these products.

Table. Organic nutrient sources for essential elements other than nitrogen.

Essential Elements	Sources
P (phosphorous)	Poultry liter, colloidal, soft and hard rock phosphate
K (potassium)	Cover crops, mined granite, greensand, basalt,
Ca, Ng, S	feldspar, langbeinite and Potassium sulfate
Micronutrients	Kelp and sea weed extracts and powders, dolomite,
(Bo, Cu, Fe, Mn,	gypsum, keiserite, langbeinite, limestone, rock
Mo, &Zn)	phosphate, and, oyster, clam and crab shells
	Liquid or powdered seaweed extract, kelp meal, rock powders

Certain meat processing industry by-products such as blood and bone meal have also been touted as organic fertilizers. However, these material also have serious food safety issues regarding their use due to the potential transmission of certain diseases to humans.

New organic farming operations and/or transition fields (those previously used for conventional crop production) may initially experience nutrient deficiencies when using organic fertilization systems until the nutritional benefits from these systems begin to become available. In these situations there are other materials approved for use. There are other growth enhancing products being used that are not fertilizers but are auxin or hormonal in nature. Kelp and other see weed extracts fall into this category. Test results, however, with many of these products have shown marginal benefits from their use.

Most conventional farming operations follow very intensive soil preparation activities such as mold board plowing, disking and bedding. Unfortunately, intensive tillage can be counter productive to an organic farming system due to its negative impact on soil organic matter content. Long-term tillage practices can reduce soil carbon 30 – 50 per cent. Consequently, in organic production where soil organic matter is the key component, conservation tillage may be a better alternative. Conservation tillage is a practice in which at least 30 per cent of the soil surface is covered by residue from a previous crop. The biggest draw back to this practice is a concern of potential weed buildup and some allopathic responses. However, if properly managed, these problems can be minimized.

11

Plant Growth Substances and their Role in Crop Production

THE NATURE OF PLANT GROWTH PROCESSES

The processes concerned in the growth of plants are the subjects of study by plant physiologists and plant biochemists. A comprehensive account of these processes is outside the scope of the present work, the special object of which is to deal with the outward and visible signs of imperfections in the plant's activities caused by faulty mineral nutrition. Nevertheless it is useful to have before us the general features of the main processes involved and to realize that the symptoms we shall be discussing later have a physiological basis, and are not direct and unchangeable signs of the specific deficiencies but result from the derangement of the complicated mechanism of the plant's vital activities.

The main processes involved in plant development may be summarized as follows:

Absorption: Intake of water and mineral elements by the root system.

Carbon assimilation or photosynthesis: Intake of carbon dioxide from the air by the leaves, and reaction of the gas with water in the leaf in the presence of the green chlorophyll to form sugar and free oxygen.

Formation of protoplasm: Protoplasm is the living material of the plant, consisting mainly of proteins, complex compounds of nitrogen built up by the plant from more simple compounds of this element.

Respiration: The combination of oxygen with various food substances synthesized by the plant, especially sugars, whereby energy is produced.

Transpiration: Loss of water from the plant, mainly from the leaves.

Translocation: The movement of materials within the plant.

Storage: Storage of reserve products in various organs and tissues.

During growth there are continuous processes of building up of complex compounds of carbon and nitrogen and breaking down of these into more simple substances, in which water and oxygen are intimately concerned. These processes together comprise plant metabolism. In the course of the metabolic processes innumerable substances are formed, such as sugars, starch, cellulose, acids, lignin, tannins, amino acids, proteins, amides etc., and many plants also produce special products, as for instance nicotine in the tobacco plant.

For the normal functioning of the above processes there must be an adequate intake of water by the plant to maintain the plant cells in a more or less turgid condition and, since water is being continuously lost at a varying rate from the plant, intake and movement within the plant tissues must be capable of ready adjustment to these changes.

As a result of metabolic activities plants develop special organs of growth and reproduction, each of which has its special characters and makes particular demands on the nutrient supplies of the plant.

With all plants there are well defined seasonal growth cycles. Thus annuals, such as cereals, begin from the seed, give rise to seedlings, which later flower, form grain and ripen off, whilst perennial deciduous trees, such as apples, pears, etc., begin growth in the spring, using stored reserves of food, form leaves, make shoots, blossom and form fruits and subsequently shed their leaves, but meanwhile pass on reserve foods to various storage organs in preparation for the next season's growth. Coincident with these growth cycles there are well defined chemical cycles of nutrient elements and elaborated products in the leaves, stems and roots, etc. It will be shown later that these cycles are of great importance in considering p-deficiency effects and in diagnosing their causes.

THE PLANT ENVIRONMENT

Nutritional problems must be considered in relation to all the conditions in which plants live, and not merely in terms of the amounts of plant nutrients contained in or added to the soil. For example, those who are accustomed to growing plants know that the temperature must not be too low or no growth may result or that if too high the plants may be injured. An optimum temperature is usually recognized and this may vary according as to whether the plant is young or old. Similarly the importance of light is well known and plants may be put in special positions to obtain a maximum supply of light at one stage and may be shaded at another.

The actual duration of the daily period of illumination also affect growth and there are plants which are classified as requiring "long day" conditions to complete their growth cycles and others as needing "short day" conditions. If the special "long" or "short" day periods are not forthcoming for the respective classes of plants requiring these, their growth cycles are abnormal and they may fail entirely to produce flowers, grain or fruit.

The humidity of the atmosphere, as distinct from the water supply in the soil, is of importance in determining the water conditions within the plant, as these are dependent on both water intake by the roots and water loss from the leaves, the latter being largely influenced by the air humidity.

Even the presence of adequate quantities of plant nutrients in the soil is no guarantee that they will be absorbed by the plant roots. It will be shown later how these may be present in forms which are not available to the plants, but even when they would be considered as being present in suitable forms for absorption, other factors may prevent this taking place. An example of this latter condition is afforded in poorly aerated soils where lack of oxygen near the roots may prevent them from actively absorbing mineral nutrients

The problems of such influences in the plant environment as those just mentioned are complicated by the fact that they do not act independently, but their effects are modified by one another. Thus the effects of light intensity or period of daylight may vary with different temperature conditions.

The requirements of plants for different nutrients may be affected by conditions of light, temperature and water supply, and by other factors of the general environment. Thus the need for nitrogen may be less under conditions of relatively low light intensity whereas the need for potash in these circumstances may be greater, these facts being of importance in growing tomatoes under glass. The effect of nitrogen in relation to light may be shown by growing a plant under normal light conditions with insufficient nitrogen, when the leaves will show the well known symptoms of nitrogen deficiency-pale green, yellow, orange and red tints. If such a plant be then shaded, the leaves will turn a darker green and growth may be visibly increased. It can be shown that the lowered light conditions result in an increase of "soluble" a breaking down of proteins, thereby rendering the nitrogen of these available for growth processes.

This interrelationship of environmental factors is well illustrated by an experiment on apple trees at Long Ashton.

Bramley's Seedling trees were grown in compost in large pots and given a small dressing of a nitrogenous fertilizer. Some of the trees were grown in a specially constructed glass house and an equal number in an adjoining wire enclosure. The trees in the enclosure showed severe symptoms of nitrogen deficiency-pale green and yellow leaves, reddish brown barks and highly coloured, red fruits. The condition was corrected by further dressings of nitrogen. In contrast, the trees under glass, where the light was of less intensity and the temperature higher, made vigorous growth, carried large, green leaves and bore large, green fruits.

Iron and zinc deficiency symptoms may be less severe under conditions of low light intensity, whilst boron deficiency effects are less severe and magnesium deficiency effects are more pronounced in wet seasons than in dry ones.

The rate of water absorption is less at lower temperatures than at higher ones and efficient intake is also dependent on good aeration. These facts may result in a water deficit within plants growing in cold, wet soils when the air temperature is high.

Soil conditions greatly complicate the problems of nutrient supplies to crops.

The raw materials needed for plant growth consist of carbon dioxide, which is obtained from the atmosphere through the stomata of the leaves, and water and the so-called mineral nutrients, which normally enter the plant through the medium of the roots.

The importance of water and carbon dioxide in the nutrition of plants will be apparent from the facts that water often comprises 80 to 90 per cent of the total weight of growing plants, and carbon and oxygen together may account for over 80 per cent of their dry matter, *i.e.*, the solid matter remaining after water is removed. As against these large amounts, the mineral nutrients, as measured by the ash content of the plants, *i.e.*, the mineral residue obtained when the organic matter is destroyed by heat, often contribute from 5 to 15 per cent of the dry matter.

It has been shown in recent years that certain organic compounds, known as "growth promoting substances" or "hormones", which occur in plants, and some of which are also present in soils and natural manures, are capable of producing marked growth responses, such as increased root growth, shoot and leaf curvatures, stimulation or suppression of buds, increased fruit setting, prevention of fruit abscission etc. They appear to perform important functions in the growth of plants. Examples of substances of this kind which can produce growth responses are 13 indole-acetic acid, 13 indole-butyric acid, phenyl acetic acid, A, naphthalene-acetamide, vitamin B_1

It is not at present clear to what extent growth substances are absorbed by plants from soils, although it has been shown that vitamin B_1, which occurs naturally in soils, can be obtained in this way.

THE MINERAL NUTRIENTS

It has been shown in numerous researches that certain elements are necessary for the healthy growth of plants. They are sometimes spoken of as essential elements and, since some are needed in relatively large quantities and others in very small amounts, the former are referred to as "major" elements and the latter as "minor" or "trace" elements, or as micro nutrients.

The terms "major" and "minor" do not refer to the relative importance of the functions of the elements in plant growth, and for this reason the term "trace" element is preferable for the latter class.

Major elements: Nitrogen, phosphorus, calcium, magnesium, potassium, sulfur.

Trace elements: Iron, manganese, boron, copper, zinc and molybdenum. (Iron occupies an intermediate position and is usually included in the major elements group. In dealing with field problems it is more convenient to group it with the trace elements.)

In addition, there are, other elements, such as sodium, chlorine and silicon, which produce beneficial effects on the growth of certain plants but which have not so far been shown to be absolutely essential to growth. The element aluminum is of general occurrence in plants, but seems to be without direct nutritional value, although aluminum sulfate is used, because of its acidifying properties, to change the colour of hydrangeas growing on alkaline soils from pink to blue, and aluminum may also exert indirect influences on nutritional processes.

Other elements often occur in plants but they are not known to serve any useful function and frequently they act as plant poisons or toxins.

The nutrient elements can only be absorbed by plants when present in certain forms: nitrogen from nitrates and ammonium salts; phosphorus from phosphates; calcium, magnesium and potassium from their salts (*e.g.*, as sulfates or chlorides, etc.); sulfur from sulfates; iron from ferrous or ferric salts (more readily from ferrous salts); manganese from manganous salts; boron from borates; copper and zinc from their salts, and molybdenum from molybdates.

There may appear to be certain exceptions to this statement in practice. For instance, nitrogen may be applied to a soil as "organic" nitrogen, as in hoof meal or urea, and sulfur may be added as the element itself, as in flowers of sulfur, ground sulfur, etc. In such conditions the added materials are, however, converted into the nitrate and sulfate forms respectively by soil organisms before being absorbed by the plants.

Further points of importance in connection with the absorption of the mineral nutrients by plants are as follows:

(a) They must be absorbed from relatively dilute solutions or the plants will be! injured or even killed.

(b) Certain of the elements slow down the absorption of others into the plant, *e.g.*, calcium slows down potassium and vice versa. The phenomenon is known as "antagonism".

(c) Healthy plants result when the nutrients are absorbed in certain relative proportions. When the proportions are suitable the nutrient medium is said to be "balanced". When ratios between nutrients are too wide, deficiency conditions are created. Thus if a high proportion of nitrogen to potassium is absorbed, the plant will suffer from potassium deficiency.

(d) Nutrients, even though present in the nutrient solution in satisfactory amounts and proportions, may not be absorbed by the plant unless the "reaction" of the solution as regards acidity and alkalinity is satisfactory. The reaction is measured in terms of the pH scale, which is merely a convenient notation for stating the conditions of acidity in the solution (strength or intensity of acidity, not total amount). The neutral point (*i.e.*, when acidity and alkalinity are equal and neutralize the effect of each other) is represented by pH 7.0; below this value the solution is acid and above it is alkaline. Many crop plants prefer a reaction slightly on the acid side-pH 6.0 to 6.5 and extreme values are in the neighbourhood of 4.0 on the acid side and 9.0 on the alkaline side.

(e) The nutrient medium must contain an adequate supply of oxygen, *i.e.*, aeration must be satisfactory.

THE FUNCTIONS OF THE MINERAL NUTRIENTS

Knowledge of the main functions of the mineral nutrients is useful in helping us to understand the effects produced by deficiencies of any one of them.

ESSENTIAL MAJOR ELEMENTS

Nitrogen. Nitrogen is a major constituent of several of the most important substances, which occur in plants. It is of outstanding importance among the essential elements in that nitrogen compounds comprise from 40 to 50 per cent of the dry matter of protoplasm, the living substance of plant cells. For this reason nitrogen is required in relatively large quantities in connection with all growth processes in plants. It follows directly from this that without an adequate supply of nitrogen appreciable growth cannot take place and that plants must remain stunted and relatively undeveloped when nitrogen is deficient.

Proteins, which are of great importance in many plant organs, *e.g.*, seeds, are compounds of nitrogen whilst chlorophyll, the green colouring matter of the leaves, also contains the element.

From this latter fact it will be apparent that when nitrogen is deficient leaves will contain relatively little chlorophyll, and will thus tend to be pale green in colour.

In addition to the above substances, numerous other organic compounds of importance in plants, such as amino acids, amides, and alkaloids, are compounds of nitrogen.

Certain compounds of nitrogen are very mobile in plants, and this enables them readily to mobilize supplies of the element at vital growing points and to transfer stored supplies to points where they are most required. Such transference is common from old tissues to young growing points when supplies of the element are short.

This mobility and re-utilization of nitrogen explains why deficiency symptoms of the element always appear first in the older parts of plants and why growing points are the last to be affected.

Phosphorus: This element, like nitrogen, is closely concerned with the vital growth processes in plants as it is a constituent of nucleic acid, and nuclei in which this occurs are essential parts of all living cells. Hence a deficiency of this element will also be expected to result in greatly restricted growth. Phosphorus is also of importance in seeds and in connection with the metabolism of fats. Compounds of phosphorus are concerned with the processes of respiration and with the efficient functioning and utilization of nitrogen. This relationship to nitrogen probably accounts for the fact that several of the symptoms of phosphorus deficiency are identical or similar to those which result from a deficiency of nitrogen, Phosphorus is also of special importance in the processes concerned in root development and the ripening of seeds and fruits.

Calcium: Calcium occurs in plants chiefly in the leaves and the amounts present in seeds and fruits are relatively low. One of its main functions is as a constituent of the cell wall, the middle lamella of which consists largely of calcium pectate. This function appears to be of fundamental importance since, if calcium is replaced by any other of the essential elements, such as magnesium or potassium, the organic materials and mineral salts in the cells are readily leached through the walls.

Other functions attributed to calcium are as follows:

It provides a base for the neutralization of organic acids; it is concerned with activities of growing points (meristems), especially with root tips; it may be of importance in nitrogen absorption.

Although a large proportion of the calcium contained in the plant may be soluble in water-as much as 60 per cent in cabbage calcium does not appear to move freely from the older to the younger parts of plants, and hence young tissues contain lower proportions of calcium than older ones. This may explain why calcium deficiency effects begin at the tips of shoots.

Magnesium: The outstanding fact about magnesium is that it is a constituent of chlorophyll, and is essential to the formation of this pigment. As a result, when magnesium is deficient, one of the symptoms commonly shown by plants is chlorosis. Magnesium is also regarded as a carrier of phosphorus in the plant, particularly in connection with the formation of seeds of high oil content, which contain the compound lecithin.

The element seems to be very mobile within the plant, and when deficient is apparently transferred from older to younger tissues where it can be re-utilized in the growth processes. This agrees with the observation that signs of magnesium deficiency invariably make their appearance first on the oldest leaves and progress systematically from them towards the youngest ones.

Potassium: Unlike all the other major elements, potassium does not enter into the composition of any of the important plant constituents, such as proteins, chlorophyll, fats and carbohydrates, concerned in plant metabolism. For this reason its role is more difficult to determine, and in spite of much study it cannot be said that the functions of potassium are clearly understood. The element is present in all parts of plants in large or fairly large proportions. It seems to be of special importance in leaves and at growing points, as these are especially rich in potassium. Probably the whole of the potassium in plants is present in soluble form, and most of it seems to be contained in the cell sap and cytoplasm.

It is outstanding among the nutrient elements for its mobility and solubility within the plant tissues, and these properties no doubt account for the ready way in which potassium can be re-utilized by young tissues when the element is in short supply.

Among the functions which have been attributed to potassium and the processes with which it may be concerned, the following may be mentioned: The formation of carbohydrates and proteins; the regulation of water conditions within the plant cell and of water loss by transpiration; as a catalyst and condensing agent of complex substances; as an accelerator of enzyme action (*e.g.*, for diastase); as contributing to photosynthesis through its radioactive properties.

It has been shown in many instances that the potassium content of plants is frequently much higher than is necessary for healthy growth, and it is generally considered that luxury (*i.e.*, unnecessary) absorption of potassium often takes place.

The great mobility of potassium in plants, its special importance for and its reutilization by young tissues, and its apparent functions as a regulator of plant processes on a large scale are in harmony with the observations that, when potassium is moderately deficient, the effects are seen first in the older tissues and progress from these towards the growing points, but, when the deficiency is acute, growing points are severely affected, and die-back and general collapse of the plants commonly occur.

Sulfur: Sulfur occurs in plants as a constituent of proteins (*e.g.*, cystine), and of certain volatile compounds such as mustard oil. It seems to be connected with chlorophyll formation, although it is not a constituent of this substance. Its functions in connection with proteins and chlorophyll doubtless account for the similarity of its deficiency effects to those due to deficiency of nitrogen.

ESSENTIAL TRACE ELEMENTS

Iron: Iron is closely concerned with chlorophyll formation but is not a constituent of it. Its role appears in this connection to be that of a catalyst. As a result of this function of iron, chlorosis is invariably an outstanding symptom when the element is deficient. Iron may also act as a catalyst, in the role of an oxygen carrier, in respiration.

A point of great importance in connection with iron is its relative immobility in plant tissues. Its mobility seems to be affected by several factors, such as the presence of manganese, potassium deficiency and high light intensity. There is evidence that the amount of chlorophyll is related to "active" (*i.e.*, readily soluble) iron in plants.

It will thus be seen that so-called iron deficiency in the plant may in fact usually mean iron immobility. Lack of mobility may also account for the fact that iron deficiency is first shown in the younger tissues.

Manganese: The functions of manganese are regarded as being closely associated with those of iron and as being concerned with chlorophyll formation. Hence, when manganese is deficient, chlorosis is a common symptom. Manganese may decrease the solubility of iron by oxidation and hence, an abundance of manganese within the plant may lead to iron deficiency and chlorosis.

Manganese is regarded as having the functions of a catalyst; its activities being specially concerned with oxidation and reduction reactions within the plant tissues.

Boron: The exact role of boron is not known, but again the evidence as for the other trace elements, suggests that, its functions are those of a catalyst or reaction regulator. Particular effects attributed to boron are as follows:

It can delay the onset of calcium deficiency effects but cannot replace calcium; it tends to keep calcium soluble; it may act as a regulator of potassium/calcium ratios, and of the absorption of nitrogen; it may be concerned with the oxidation-reduction equilibrium in cells.

Such functions as the above accord with the results which follow from a deficiency of the element, when growth processes show sudden collapse and drastic derangements of metabolism occur.

Zinc and Copper: Although specific functions have not been determined for these elements, here again the evidence points to their roles as catalysts and regulators. Deficiencies of both are associated with chlorosis and a serious general collapse of vital growth processes.

Since catalysts are not used up in the chemical reactions which they promote, we can understand how it comes about that quite small or even minute quantities of the "trace elements", iron, manganese, boron, zinc and copper, may nevertheless be essential to the plant's health and growth.

ELEMENTS SOMETIMES BENEFICIAL

Sodium: As sodium is not strictly an essential element it cannot be expected to have a specific role in the metabolic activities of plants.

Where sodium produces significant effects it is often regarded as a conserver of potassium and as being able partly to replace that element in its role. In no instance, however, has it been shown that sodium can wholly replace potassium where the latter is acutely deficient. In such circumstances, sodium is ineffective as a substitute for potassium, even for sodium-loving plants, such as sugar beet, mangold and barley. Sodium seems to affect the water relations of plants and often enables sugar beet and other crops to withstand drought conditions which would otherwise produce severe adverse effects.

Chlorine: The evidence of the role of chlorine in plants is somewhat contradictory, and no general statement can be made. In tobacco it has been shown to increase the water content of the tissues and to affect carbohydrate metabolism, leading to an accumulation of starch in the leaves. The element is present in plants as chloride, and is wholly soluble.

TOXIC EFFECTS OF MINERAL ELEMENTS

Toxic effects in plants may be produced both by essential nutrient elements and non-essential elements.

In the first class, the major nutrients are much less toxic than the trace elements. Indeed, for the major nutrients, there exists a fair safety margin for excess or "luxury" consumption, but for the trace elements the margin is very narrow. Similar conditions exist in relation to non-essential elements thus some plants will tolerate fairly large amounts of elements such as sodium or chlorine, but are injured by relatively small amounts of elements like arsenic or chromium.

Two types of injury may occur: (1) An excess of one element may lead to a deficiency of another which ultimately results in a deranged metabolism, *e.g.*, excess nitrogen or excess phosphorus may result in insufficient potassium and excess potassium may lead to deficiency of magnesium or calcium. This type of injury applies particularly to essential nutrients. (2) The presence of an element may directly injure the protoplasm and bring about the speedy death of the plant.

SIGNS OF FAULTY NUTRITION IN PLANTS

When plants are grown in unsuitable environments, including conditions of faulty mineral nutrition, they react to the particular defects in more or less specific ways. Thus, if light is insufficient, the green colouring matter of the leaf will be lacking and the leaves may be almost white (chlorotic), and the plant may be very spindly and "drawn" in appearance; if the temperature is too high, the growth may be rank and soft; if water is insufficient, growth may be restricted, the tissues woody and the green of the leaves show a bluish tint. Again, deficiencies and excesses of the individual elements produce characteristic effects on various organs of plants: foliage characters, including colour, density, size and shape of leaves; stem characters, such as thickness, colour and length of inter-nodes; root characters, such as colour, amount of fibre, abnormal thickening; blossom characters, including amount and time of opening of the flowers; fruit characters, such as size, colour, hardness and flavour.

Ability to recognize these particular effects forms the basis of the visual method of diagnosing plant deficiencies. Many of them can be readily learned and applied by practical farmers. Indeed, for many years progressive fruit growers in this country have used the leaf symptoms of deficiencies of nitrogen and potassium, and more lately of magnesium, as the main guide to manuring their fruit trees and bushes with these elements.

PLANT GROWTH SUBSTANCES IN CROP PRODUCTION: A REVIEW

Nutrients are important and crucial elements, which are required for the plant for its growth and development. The translocation of photosynthates from source to sink is very important for the development of economic part. Plant growth regulators are chemical substances and when applied in small amounts, they bring rapid changes in the phenotypes of the plant and also influence the plant growth, right from seed germination to senescence either by enhancing or by stimulating the natural growth regulatory system. Plant growth regulators are known to enhance the source-sink relationship and stimulate the translocation of photo-assimilates thereby helping in effective flower formation, fruit and seed development and ultimately enhance productivity of the crops.

Growth regulators can improve the physiological efficiency including photosynthetic ability and can enhance effective partitioning of the accumulates from source and sink in the field crops. Foliar application of growth regulators and chemicals at the flowering stage may improve the physiological efficiency and may play a significant role in raising the productivity of the crop.

Importance of Salicylic Acid (SA): Salicylic acid belongs to an extra-ordinary diverse group of plant phenolics. Salicylic acid is ubiquitously found in all plant species and the highest levels are observed in the inflorescence of thermogenic plants and in plants infected with necrotizing pathogen.

Salicylic acid is ortho-hydroxybenzoic acid and is a secondary metabolite acting as analogues of growth regulating substances. It helps in protection of nucleic acids and prevention of protein degradation. The SA is also known to induce many genes coding for pathogenesis-related proteins in response to biotic and abiotic stresses.

Foliar application of SA increased the IAA content in broad bean leaves. Foliar application of SA exerted a significant effect on plant growth metabolism when applied at physiological concentration, and thus acted as one of the plant growth regulating substances.

Effect of salicylic acid on growth parameters: Jeyakumar *et al.* (2008) reported that application of salicylic acid (125 ppm) increased thedry matter production (21.6 g plant^{-1}) in black gram. Nagasubramaniam *et al.* (2007) reported that foliar application of salicylic acid (100 ppm) on baby corn increased the plant height, leaf area, crop growth rate and total dry matter production. Khan *et al.* (2003) reported that application of salicylic acid, acetyl salicylic acid and geninic acid enhanced the photosynthetic rate, stomatal conductance and transpiration rate in soybean. Sujatha (2001) reported that foliar application of salicylic acid (100 ppm) on green gram at 75 DAS increased the plant height (50.4 cm), root length (16.9 cm), number of leaves (18.4) and LAI (1.30).

Effect of salicylic acid on yield and yield parameters: Jeyakumar *et al.* (2008) reported that application of salicylic acid (125 ppm) in black gram increased the seed yield (855 kg ha^{-1}). Salicylic acid sprayed on mungbean significantly increased the pod number $plant^{-1}$and yield (Singh and Kaur, 1981). Nagasubramaniam *et al.* (2007) revealed that the cob yield of baby corn was the highest (7703.4 kg ha^{-1}) by the foliar application of salicylic acid (100 ppm).

Sujatha (2001) reported that foliar application of salicylic acid (100 ppm) on green gram at 75 DAS increased number of pods $plant^{-1}$, number of seeds pod^{-1}, seed weight $plant^{-1}$, 100 seed weight and grain yield (840 kg ha^{-1}). In green gram, foliar application of salicylic acid at branching, flower bud initiation stages increased the number of flowers, pods and seeds $plant^{-1}$ and seed yield.

Effect of salicylic acid on quality parameters: Jeyakumar *et al.* (2008) reported that the highest seed protein content in black gram was recorded (24.5 per cent) by foliar application of salicylic acid (125 ppm). Hussein *et al.* (2007) indicated that in maize hybrid, all determinateamino acids concentration such as tyrosine (31.16 per cent), lysine (22.30 per cent), aginine (20.68 per cent), alinine (2.65 per cent), lucine (2.78 per cent) except methionine were increased with the application of salicylic acid (200 ppm). Proline concentration increased when using salicylic acid as foliar application under salt stress condition. Sujatha (2001) revealed that foliar application of Salicylic acid (100 ppm) on green gram at 75 DAS increased seed protein (23.98 per cent) and soluble protein (12.9 per cent). Kalpana (1997) found that foliar spray of salicylic acid increased the soluble protein in rice.

Effect of salicylic acid on nutrient uptake: Gunes *et al.* (2007) at Turkey demonstrated that exogenously applied salicylic acid increased the plant growth in maize significantly both in saline and non-saline conditions. This may be related to its inhibiting effect on Cl^{-1} and Na^{+1} and improving the uptake of N, Mg, Fe, Mn and Cu due to its effect on lipid peroxidation.

Importance of boron: Boron is one of the important micronutrients, which has basic role in stabilizing certain constituents of cell wall and plasma membrane, enhancement of cell division, tissue differentiation and metabolism of nucleic acid, carbohydrate, protein, auxin and phenols. Though the symptoms of boron are very striking, its role in plant metabolism is uncertain. Boron has been found to be involved in the carbohydrate transport system within the plant. Boron deficiency causes delay in tassel formation, decrease in pollen size, pollen germination and pollen production capacity of maize. Among the micro nutrients, boron is stated to influence many growth parameters and filling up of seeds.

Effect of boron on growth parameters: Vishwakarma *et al.* (2008) revealed that in groundnut maximum plant height (60.33 cm) and number of branches (5.83 plant^{-1}) were recorded with application of borax as soil application. Dixit and Elamathi (2007) reported that foliar application of boron (0.2 per cent) in green gram increased the plant height (32.26 cm), number of nodules plant^{-1} (30.8) and dry weight plant^{-1} (12.9 g).

Garg *et al.* (1979) found that germination capability, size and fertility of pollen grains of rice were considerably improved as a result of boron application at 2.5 ppm concentration. They also reported that inclusion of boron in the nutrient solution with optimum level induced some stimulatory effects on the pollen vitality and there by improved the grain yield of rice.

Effect of boron on yield and yield parameters: The highest number of pods plant^{-1} (26.66), pod and kernel yield (25.46 and 16.34 q ha^{-1}, respectively) were recorded with application of borax as soil application in groundnut. The maize crop sprayed with boric acid @ 0.3 per cent produced higher grain yield (6444 kg ha^{-1}) than no spray treatment.

Dixit and Elamathi (2007) reported that foliar application of boron (0.2 per cent) in green gram increased the number of pods plant^{-1} (18.1), 1000 seed weight (28.7 g), grain yield (7.53 q ha^{-1}) and haulm yield (30.0 q ha^{-1}). Susheela (1996) revealed that foliar spraying of borax at 0.2 per cent increased percentage of filled seeds in sunflower. This might be due to better pollen germination, increased pollen viability and fertilizing capacity.

Khan *et al.* (1990) found that dusting of borax at 2 kg ha^{-1} on sunflower heads during seed filling stage significantly increased the seed yield by 87 per cent, test weight by 60 per cent, seed filling by 25 per cent and head diameter by 21 per cent compared to control. They also found dusting borax on ear heads gave the highest germination per cent of 96 and vigour index of 2162 compared to control (84 per cent germination and 1263 vigour index, respectively).

Fernandez (1990) reported that the requirement of boron was maximum during flowering and the plant was therefore most susceptible to limitations and so the content of the 5th and 6th leaves under the flowering head should be maintained at 20 and 40 ppm, respectively.Sarkar and Sasmal (1989) found that applying 0.5 kg ha^{-1} boron to sunflower grown in *rabi* (winter) season increased the capitulum diameter, number of filled seeds head^{-1}, 1000 seed weight and seed yield.

Randhawa and Nayyar (1982) reported that application of boron was found to be beneficial in increasing the yield of wheat and paddy.Satyanaryana *et al.* (1997) reported that application of boron at 2 and 4 kg ha^{-1} increased the yield of sunflower by 38 and 59 per cent, respectively over control. Rao and Vidyasagar (1981a) reported that foliar application of boron increased the plant height, stem girth, dry matter accumulation, total seed number per head, number and weight of filled seed per head, test weight and seed oil content.

Effect of boron on quality parameters: Application of boron either alone or in combination with FYM and sulphur helped in improving the quality of groundnut. Sekar *et al.* (2009) recorded higher starch content in hybrid maize sprayed with boric acid @ 0.3 per cent.

Effect of boron on nutrient uptake: Jain *et al.* (2007) revealed that Rhizobium inoculation along with zinc and boron in green gram gave better response in terms of N uptake as compared to Rhizobium alone. Ateeque and Malewar (1992) reported that uptake of N, P and boron was the highest when fertilized with boronated super phosphate as compared to SSP and DAP.

Rao and Vidyasagar (1981b) reported that boron when applied to soil resulted in increased N content of leaf, K content of stem and P content of petiole during early vegetative stage, N and P contents of leaves during flowering stage and P contents of leaves and stems, and N and K contents of petioles at harvest. Boron when sprayed on foliage helped to increase the mineral composition of leaves, stems and petioles during different growth stages of sunflower. They also noticed that foliar sprayed boron influenced the absorption and preferential translocation of mineral elements into different parts of the plant.

Effect of boron on economics: Dixit and Elamathi (2007) reported that foliar application of boron (0.2 per cent) in green gram increased the B: C ratio (1.54). Susheela (1996) revealed that net return and B: C ratio was higher with 60 kg P_2O_5 ha $^{-1}$ combined with of 2 per cent DAP and 0.2 per cent borax spray in sunflower.

Importance of panchagavya: In 1950, James F. Martin of USA made a liquid catalyst (living water) from milking cow, using dung, sea water and yeast and it was claimed that it was capable of greening degraded land. Cows ghee had been used in ancient and medieval times for managing seedling health. The ghee contains vitamin A, vitamin B, calcium, fat and also glycosides, which protects cut wounds from infection. Cows curd is rich in microbes (*Lactobacillus*) that are responsible for fermentation.

Panchagavya has got reference in the scripts of Vedas (divine scripts of Indian wisdom) and Vrkshayurveda (Vrskha means plant and ayurveda means health system). The texts on Vrskhayurveda as systematization of the practices that the farmers followed at field level, placed in a theoretical frame work and it defined certain plant growth stimulants; among them Panchagavya was an important one that enhanced the biological efficiency of crop plants and quality of fruits and vegetables. The positive effect of panchagavya on growth and productivity of crops has been reviewed and documented by Somasundaram and Amanullah (2007). Effect of panchagavya spray on growth parameters: Mohanalakshmi and Vadivel (2008) revealed that ashwagandha plant sprayed with panchagavya (3 per cent) produced higher number of leaves plant^{-1} (670). Vennila and Jayanthi (2008) revealed that application of 100 per cent recommended dose of fertilizer along with panchagavya spray (2 per cent) significantly increased the okra plant height (131.7 cm) and dry matter production (5.90 g plant^{-1}).

Biogas slurry with panchagavya combination is adjudged as the best organic nutrition practice for sustainability of maize-sunflower-green gram system by its overall performance on growth, productivity, quality of crops, soil health and economics.

Panchagavya was tested for different crops such as turmeric, paddy, onion, gingely, sugarcane, banana, vegetables and curry leaf and it was found that it enhanced the growth, vigour of crops, resistance to pest and diseases and improvement of keeping quality of vegetables and fruits.

Xu (2001) reported that Effective Micro Organism (EMO) cultures could synthesize phytohormones *i.e.*, auxins and other growth regulators that stimulated maize plant growth and they contained proactive substances that could significantly affect leaf stomatal response in maize. Leaf stomata of the EMO treated maize opened more rapidly than water treated control plants and when leaves were subjected to dehydration, the stomata closed more slowly (*i.e.*, remained open longer) thus showed that, EMO contained bioactive substances that could have significantly affected leaf stomata response and led to increased LAI.

Effect of panchagavya spray on yield and yield attributes: Mohanalakshmi and Vadivel (2008) revealed that application of poultry manure (5 t ha^{-1}) + panchgavya (3 per cent) in aswagandha exhibited significantly superior performance by registering the highest root yield of 1354.50 kg ha^{-1}. Vennila and Jayanthi (2008) revealed that application of 100 per cent recommended dose of fertilizer along with panchagavya spray (2 per cent) significantly increased the number of fruits plant^{-1} (21.8), fruit weight g fruit^{-1} (13.9) and fruit yield q ha^{-1}(12.7) of okra.

Swaminathan *et al.* (2007) concluded that application of panchagavya at 3 per cent as foliar spray on 15, 25, and 40 DAS on black gram recorded the highest grain yield of 1195 kg ha^{-1}. Kanimozhi (2003) revealed that application of *panchagavya* at 4 per cent spray was found to be superior in respect of root yield (12.4 kg/plot) compared to control (5.23 kg plot^{-1}) in *Coleus forskohili*.

Foliar spray of *Panchagavya* @ 3 per cent on 15, 25, 40 and 50 DAS with no fertilizers was the most effective low cost technology in terms of grain yield of greengram (Somasundaram *et al.*, 2003). *Panchagavya* and vermicompost combination has given the highest pod yield of french bean variety Ooty 2, which was 36 per cent higher than the conventional method. The treatment combinations of poultry manure + neem cake + panchagavya along with increased dose of fertilizers increased the stick yield of moringa.

Balasubramanian *et al.* (2001) reported that dipping of rice seedlings in *panchagavya* before transplanting enhanced the growth and yield.

In panchagavya, Effective Micro Organisms (EMO) were the mixed culture of naturally occurring, beneficial microbes (mostly lactic acidbacteria (Lactobacillus), yeast (Saccharomyces), actinomyces (Streptomyces), photosynthetic bacteria (Rhodopsuedomonas) and certain fungi (*Aspergillus*)) and that improved the soil quality and growth and yield of sweet corn, which was equal to or higher than what was obtained from chemical fertilizers.

In jasmine, spraying two rounds of panchagavya, one before the flower initiation and another during bud setting phase ensured continuous flowering and in annual moringa sprayings doubled the stick yield besides giving resistance to pests and diseases. Panchagavy*a* sprayed on 25 and 40 DAS advanced the paddy harvest by 10 days.

Effect of panchagavya spray on quality parameters: Vennila and Jayanthi (2008) revealed that application of 100 per cent recommended dose of fertilizer along with panchagavya spray (2 per cent) in okra resulted in higher crude protein, ascorbic acid and Barletts index. The presence of auxin in panchagavya controls the water regulation in developing fruits. Regular and uniform water supply to the developing fruits resulted in increased ascorbic acid content, Barletts index and crude protein content.

Yadav and Lourduraj (2006b) revealed that foliar spray of panchagavya and organic manures recorded better cooking qualities and physical characteristics of rice as well as higher sensory score as compared to recommended N, P and K through fertilizers.

Effect of panchagavya spray on nutrient uptake: Presence of macro (N. P, K and Ca) and micro (Zn, Fe, Cu, Mn) nutrients besides total reducing sugars (glucose) were observed in panchagavya. Chemolithotrops and autotropic nitrifiers (ammonifers and nitrifers) present in panchagavya which colonize in the leaves increase the ammonia uptake and enhance the total N supply.

Beaulah (2001) reported that the secondary and micronutrients (Ca, S and Fe) and macronutrients (NPK) contents of leaves and pods of annual moringa were superior under poultry manure + neem cake + panchagavya treatments. Higher nutrient uptake and nutrient use efficiency in both main and ratoon crops of annual moringa were also observed. Similarly, the quality parameters *viz.*, crude fibres, protein, ascorbic acid, carotene content and shelf life were also higher under organic manure applied with panchagavy*a* spray.

Influence of panchagavya spray on soil fertility: Microbial flora of soil play an important role in soil health. The microorganisms present in the rhizospheres environment around the roots influence the plant growth and crop yield. The beneficial microorganisms from panchagavya and their establishment in the soil improved the sustainability of agriculture. Effect of panchagavya spray on economics: Swaminathan *et al.* (2007) reported that application of panchagavya at 3 per cent as foliar spray on 15, 25, and 40 DAS on black gram under irrigated condition recorded the highest net return (₹ 28544 ha^{-1}) and B:C ratio (4.91).

Application of 50 per cent N through composted poultry manure + 50 per cent N through green leaf manure along with panchagavya spray recorded higher net returns (₹ 17822 ha^{-1}), followed by recommended NPK through panchagavya spray (₹ 15586 ha^{-1}) in rice.

Importance of Pink Pigmented Facultative Methylotroophic bacteria (PPFMs): Pink Pigmented Facultative Methylotrophs (PPFMs) belonging to the genus *Methylobacterium* are persistently present in the rhizosphere and phyllosphere regions of plants and even on the surface of the seeds of various plants. They were revealed on leaves of almost all plants.

Methylotrophs have received a great deal recently due to their importance in the biosphere and their potential commercial applications.Holland (1997) reported that the invention provided a method for increasing productivity of a plant by spraying PPFM on a plant. The invention also related to increasing productivity of a plant under stress by applying PPFM to a plant and subsequently applying an aqueous solution-containing methanol to the plant.

Methylobacterium have some sort of beneficial interaction with plants. Pink pigmented facultative methylotrophs (PPFMs) isolated from the liverwort stimulated growth of the liverwort (*Scapania nemorosa)* and *Streptocarpus prolixus*(flowering plant) in tissue cultures and a positive commensal interaction was proposed. Effect of PPFMs on germinating seeds has been suggested to be mediated by cytokinin or other plant growth promoting substances and a possible connection between bacterial and plant metabolism has also been suggested.

Effect of PPFM spray on growth attributes: Under the pot culture condition in blackgram variety CO_2, the plant height was significantly increased (37.28 cm) by foliar application of urea and phyllosphere spraying of *Methylotrophs extorquees* PPFMs-Vm-11, which was superior to uninoculated control (30.28 cm $plant^{-1}$).

Madhaiyan, 2003 reported that inoculation of *Methylobacterium* and methanol spray significantly increased the plant height and dry matterproduction of cotton than uninoculated control. The inoculation of *M. extorquens* PPFM-GO-71 by phyllosphere spray with methonal had significantly increased the plant height (84.3 cm $plant^{-1}$) and dry matter production (15.12 g $hill^{-1}$) than all the other treatments.

Sundaram *et al.* (2002) stated that PPFMs influenced seed germination and seedling growth by producing plant growth regulators zeatin and related cytokinin. Due to the production of plant hormone and vitamins by pink pigmented facultative methylotrophic bacteria colonizing on the phyllosphere regions of maize, PPFM enhanced the seedling vigour in maize. Inoculation of *Methylobacterium* sp. PPFMs-Ah recorded maximum plant growth of groundnut at harvest (57.0 cm) which was followed by *Methylobacterium* sp. PPFMs-Os (52.7 cm) and*M. extorquens* AMI (51.1 cm).

Effect of ppfm spray on yield and yield attributes: Radhika *et al.* (2008a) revealed that the highest maize cob yield (7785 kg ha^{-1}) was obtained from foliar application of PPFMs at 5 L ha^{-1}. The highest cob yield was due to significant increase in length, diameter and weight of cob and corn. Exogenous methanol of plant stimulated the growth of *methylobacterium* sp, which provided plants with cytokinins and auxin which in turn enhanced plant development and ultimately increased the yield. This might be due to supply of cytokinin by PPFMs. Cytokinin helps in promotion of cell division, delaying of senescence, counteracting apical bud dominance, translocation of assimilates and thereby improve the yield potential of plants.

Madhaiyan, 2003 observed that application of PPFMs as foliar spray significantly increased the boll numbers, boll weight and kapas yield of cotton. Foliar application of urea and phyllosphere spraying of *M. extorquens* PPFMs-Vm-11 significantly increased the seed weight of black gram (7.52 g $plant^{-1}$) over uninoculated control (5.12 g $plant^{-1}$).

The highest boll and seed weight of cotton was obtained with phyllosphere spraying *M. extorquens* AMI with methanol spray followed by seed imbibitions of *M. extorquens*, PPFMS-GO-71 reported by. Reddy (2002) revealed that the inoculation of PPFMs isolate *Methylobacterium* sp. PPFMs-Ah recorded higher number of nodules at 60 (30.33), 90 (52.32) and 120 DAS (61.22) than other treatments in groundnut. Reddy (2002) revealed that *Methylobacterium* sp. PPFMs-Ah recorded maximum pod weight of groundnut (10.60 g) by seed imbibition method, where as *Methylobacterium* sp. PPFMs-Os recorded 8.98 g and M. extorquenes AMI 8.86 g.

Effect of PPFM spray on quality of crops: Radhika *et al.* (2008b) concluded that application of pink pigmented facultative methylotrophs at 5 L ha^{-1} increased the productivity as well as quality parameters of baby corn. Madhaiyan *et al.* (2005) revealed that inoculation of PPFM resulted in significant increase in sugar cane plant growth, cane yield and sugar quality.

12

Nanoclay Clay Polymer Composite and their Role in Crop Production

BIOPLASTIC CONTAINER CROPPING SYSTEMS

The goal of this project is to develop and evaluate commercially feasible bioplastic cropping containers that will fulfill all of the functions of petroleum-based plastic containers without their drawbacks. The biopolymers are being prepared from numerous biorenewable proteins, natural oils, carbohydrates, and composites of these and other natural materials. We are also considering the use of recently developed natural oil-based coatings to enhance water resistance and longevity of containers constructed from proteins, natural oils, carbohydrates, and other composites.

Fig. Bioplastic pots developed for this project.

Thirty-four different types of biorenewable polymers and composites are being screened to evaluate their potential for use in container crop production and to determine the rate of biodegradation in soil under landscape conditions. These containers are made from a variety of biobased materials including soy protein, wheat starch, two commercially available plant starch-based plastics PHA (polyhydroxyalkanoates) and PLA (polylactic acid), and natural fibres (wood, paper, and coconut husk). PHA and PLA were also blended with dried distillers grains with solubles (DDGS), corn stover, and nano-clay to form biocomposite materials.

Fig. Greenhouse trials underway for biobased plant container project

Plants from the trial are being transplanted in a garden plot at the Horticulture Research Station to assess the long-term growth and health of plants produced in biobased containers. The landscape trial will allow us to evaluate the biodegradation rate of containers and to quantify the plant growth benefit from containers as they degrade in the soil near the plant roots. In additional trials, two species of plants are grown in containers for eight weeks, as we look more closely at water use efficiency and the fertilizer effect provided by container degradation during greenhouse production.

FIBRE SPINNING

Structure and property development during fibre formation has been a major topic of my research for the past ten years. In addition to my work experience in a fibre company producing Nylon, Polyethylene Terephthalate (PET) and Acrylic fibres, and a carpet company spinning fibres from post-consumer recycled PET, I was actively involved in research on high speed melt-spun fibres from polyesters and their blends and copolymers. During my dissertation research on carbon fibres, involvement in wet spinning of acrylic fibres was also significant.

One of my masters student (Yi Xiang Zhang) completed research on melt blowing of recycled PET, the modification of the PET fibres during the process by changing process conditions, and blending with other compatible polymers. Dr. Vasanthakumar, who completed his doctoral research under my guidance worked on improving the dimensional stability of PET fibres by incorporating nucleating additives.

Dr. Dong Zhang did his PhD research on structure and property development of homopolymer and copolymer PP in a spunbonding process. In the context of spunbonding, effect of various additives such as softeners and nanoclay on processing and on the structure and properties of the resulting fibres are being investigated.

During the Summer of 1997, research was done in the area of melt spinning of polyetherimide to produce hollow fibres for special applications in collaboration with the Phillips Laboratory, Edwards AFB, CA. Current research in this area included fibres from biodegradable and bioabsorbable polymers and high temperature polymers.

BIODEGRADABLE NON-WOVENS AND COMPOSITES

For the past seven years research is being done in the area of cotton-based Non-wovens fro biodegradable applications. The idea has been to produce all-cellulose and cotton/biodegradable binder fibre Non-wovens by environmentally safe techniques. This work was supported by Cotton, Incorporated, Raleigh, NC and Tennessee Agricultural Experimental Station. This research is being extended to study the fabrication of compostable composites from all biodegradable fibres by thermoplastic molding. This involves the use of natural lingo-cellulosic fibres such as cotton/kenaf/flax with a suitable thermoplastic biodegradable binder fibre. These products are targeted for applications in automobiles, building and appliances as thermal and acoustical insulators. The collaborators in this research are, Prof. K. E. Duckett, Dr. H. Suh, Miss. X. Gao, Dr. H. Rong, M. G. Kamath, D. V. Parikh and D. Mueller. Currently research is done to understand the structure and properties of biodegradable, bioabsorbable and flushable Non-wovens.

COMPOSITE NON-WOVENS

While non-woven composites have been in use for quite some time, the future for these fabrics is more promising. In applications such as health care, engineered non-wovens seem to be the material of choice due to greater concern of spread of deadly diseases like AIDS. The research seems to indicate that use of films as a barrier to Virus is essential, making the conventional SMS type composites not sufficient to meet the stringent demands. While the use of films improves barrier properties, it affects the comfort and aesthetics of the fabrics negatively, making the use of such products extremely difficult. It is a challenge to engineer the combination of barrier to virus and good comfort in such products. Use of perforated film and non-wovens in composites and the understanding of their role in contributing to required properties are important to be able to design appropriate fabrics. Investigations will be done with suitable combinations of films and non-wovens to design fabrics that are appropriate for the required applications, and also are easily disposable.

EXTRUSION COATING

Non-wovens are becoming more and more important as coating substrates for many applications. Today's non-wovens are considered engineered fabrics that can be produced with a range of properties desired in substrates for coated textiles. As the technology has emerged, engineered coated fabrics are being used as camouflage nets, thermo chromic fabrics, protective clothing for toxic chemicals etc. Metallized coatings are finding use in EMI-RFI shielding and radar responsive materials. Many of the coated fabrics are excellent barrier systems that have great potential in protection against terrorist threats like the recent Anthrax attacks. In addition to providing the necessary barrier with selective permeability, these composite fabrics should have: high strength, dimensional stability and resistance to deformation; retention of mechanical property in a widely ranging temperature conditions; resistance to water, sunlight, and atmospheric pollutants. Research focus is to understand the coating process with different polymers and additives on textile substrates, which will help in engineering the proper structure for appropriate applications.

NANOFIBRES

Nanoscience has made tremendous advances in recent years. Development of novel structures and processes, and novel experimental and theoretical tools provides new opportunities to incorporate these technological breakthroughs into products that can have broad positive impact on our lives. Still it is often a challenge to produce such systems economically and on a large scale, limiting the potential impact of nanotechnology on the materials industry. Many of the envisioned products of nanotechnology cannot be produced by the large-scale micro fabrication techniques currently in use. In my recent work (with eSpin, Inc., Chattanooga, TN), commercially practiced processes are being used for the incorporation of or generation of nanostructures during or after a typical fabrication process. There is increasing interest on the part of many companies to adapt nanoscience to manufacturing processes. With such an approach, one should be able to produce novel structures with unique properties, whereby the molecular architecture is controlled at the nanoscale, at commercially viable rates. Another recent research topic (supported by NASA) is carbon nanotube reinforced nanofibres ribbons.

NANOCLAY AND NANO-METAL OXIDE INCORPORATED FIBRES AND FABRICS

The effect of incorporating nanoclay in polypropylene spunbond fabrics was investigated (with Kamath and Hegde). It was seen that blending about one per cent of nanoclay increases the tensile strength and modulus. What is unique is that the observed increase in strength is not accompanied by decrease in breaking elongation. Also, with proper compounding methods employed, there is no need to change any machinery or processing to achieve positive results. Current work involves the evaluation of techniques to disperse and exfoliate nanoclay in different polymers.

Nanophase Mn (VII) oxide (Agent M-7-O) incorporated non-woven fabrics are being developed for military and civilian applications (Dr. Raj Vempati of ChK Group, Inc. Dallas, TX). Agent M-7-O is an environmentally friendly, stable and a strong Lewis acid oxidant. This is a fast-moving, low-cost technology capable of destroying chemical warfare agents, as well as toxic industrial chemicals. The key benefits of this product are that, they can be safely transported, stored and handled, there is no risk to the environment, and they can be easily manufactured in different forms. Other variations of MnO are also being investigated. The focus of this research has been to understand and control the distribution of nanoparticles within the fibre structure and their activity for performance.

NANOTECHNOLOGY IN AGRI-FOOD PRODUCTION: AN OVERVIEW

The practice of agriculture also known as "farming" is the process of producing food, feed, fibre, and many other desired products by the cultivation of certain plants and the raising of livestock. Agriculture is the backbone of most developing countries and it provides food for humans, directly and indirectly. The world's population will grow to an estimated 8 billion people by 2 025 and 9 billion by 2 050, and it is widely recognized that global agricultural productivity must increase to feed a rapidly growing world population.

The agri-food production is of vital importance, as it has been one of the primary drivers of economy. In addition, it can offer routes to value-added crops. Agricultural practices are often in the public eye because climate change, energy and resource constraints, and rapidly growing global population are placing unprecedented pressure on food and water resources. The Food and Agriculture Organization of the United Nations predicts that annual meat production of 200 million tons will be required by 2 050 to respond to the food needs brought about by increasing global population,1 and this predicted increasing demand for meat puts further pressure on agricultural land because farmers need to grow crops to produce animal feed.

Land for food crops also faces increasing competition from the need for crops for other purposes – such as the production of biofuels and pharmaceuticals. Thus food production capacity is faced with many challenges, which include a falling ratio of arable land to population. Agriculture as a source of food is becoming increasingly important in a world of diminishing resources and an ever-increasing global population.2 Given the increasing world population, it is necessary to use the modern technologies such as nanotechnology and nanobiotechnology in agricultural and food sciences. Nanotechnology has a tremendous potential to revolutionize agriculture and allied fields, including aquaculture and fisheries. Nanoagriculture focuses currently on target farming that involves the use of nanosized particles with unique properties to boost crop and livestock productivity.

Agri-food nanotechnology is multidisciplinary in nature. Nanotechnology application to the agriculture and food sectors is relatively recent compared with its use in drug delivery and pharmaceuticals. Nanotechnology has the potential to protect plants, monitor plant growth, detect plant and animal diseases, increase global food production, enhance food quality, and reduce waste for "sustainable intensifcation". Food and agricultural production are among the most important fields of nanotechnology application.

NANOTECHNOLOGY AND NANOMATERIALS

Nanoscale refers to size dimensions typically between approximately 1–100 nm (or more appropriately, 0.2–100.0 nm) because it is at this scale that the properties of materials differ with respect to their physical, chemical, and biological properties from those at a larger scale. A single nanometer (nm) is 1 billionth of a meter. Nanotechnology refers to the understanding and control of matter at nanoscale, where a unique phenomenon enables novel applications. However, limiting size in nanotechnology to the 1–100 nm range excludes numerous materials and devices, especially in the pharmaceutical and agricultural fields, and some experts caution against a rigid definition based on a sub-100 nm size. Any form of a material that has one or more dimensions in the nanoscale is known as nanomaterial.

According to another definition, "nanomaterial" means a natural, incidental, or manufactured material containing particles in an unbound state or as an aggregate or as an agglomerate and where, for 50 per cent or more of the particles in the number size distribution, one or more external dimension is in the range 1–100 nm. Materials that have one dimension in the nanoscale (and are extended in the other two dimensions) are layers, such as graphene, thin films, or surface coatings. Materials that are nanoscale in two dimensions (and extended in one dimension) include nanowire and nanotube.

Materials that are nanoscale in three dimensions are particles, for example precipitates, dendrimers, fullerenes, colloids, and tiny particles of semiconductor materials (quantum dots). Nanomaterial-specific properties derive mainly from their increased relative surface area and quantum effects. Any material that is intentionally produced in the nanoscale to have specific properties or a specific composition is called a manufactured/engineered nanomaterial. Such engineered nanomaterials have different properties when compared with their conventional counterparts. A nanoparticle is a discrete entity that has all three dimensions in the nanoscale..

Biological natural nanoparticles

Biological naturally occurring nanoparticles (nanoclay, tomato carotenoid lycopene, many chemicals derived from soil organic matter, lipoproteins, exosomes, magnetosomes, viruses, ferritin) have diverse structures with wide-ranging biological roles. Biological nanoparticles are often biocompatible and have reproducible structure. Potential biomedical applications of natural and modified biological nanoparticles have been reported.

Animals use nanotechnology, where nanostructures help animals climb, slither, camouflage, flirt, and thrive. A good example is the ordered hexagonal packed array of structures in the wings of cicadas (for instance,*Psaltoda claripennis* Ashton) and termites (for example, family *Rhinotermitidae*). Studying nanostructured nipple arrays of moth eye facets helps to design better thin-film solar cells. A combination of three functions in one biological nanostructure (antiadhesive properties of insect ommatidia grating in addition to their widely accepted antireflective properties and ability to reduce glare to predators) can be exploited for the development of industrial multifunctional surfaces capable of enhancing light harvesting while reducing light reflection and adhesion. Butterfly wings contain nanostructures that give rise to optical effects such as iridescence, the effect of changing colour when viewed from different angles. The tokay gecko uses nanotechnology to stick itself to trees, walls, windows, and even ceilings. Mimicking the agile gecko, researchers have created synthetic "gecko tape" with four times the sticking power of the real thing. It is well known that insects possess ferromagnetic resonance which is temperature dependent and that magnetic nanoparticles in social insects act as geomagnetic sensors.

Nanotechnology promises to improve current agriculture practices through the enhancement of management and conservation of inputs in crops, animal production, and fisheries. In recent years, the food industry has made great progress in areas such as the improvement of new packaging products, the development of new functional products, transport and controlled release of bioactive substances, detecting of pathogens by using nanosensors and indicators, and purification of water through the use of nanoparticles. The potential for improving the effectiveness of agricultural active ingredients using nanosized particles, including functionalized nanocapsules, has been reported. Agricultural applications also include i) nanotechnology-enabled delivery of agriculture chemicals, ii) field-sensing systems to monitor the environmental stresses and crop conditions, and iii) improvement of plant traits against environmental stress and diseases.

NANOAGROCHEMICALS

Pesticides are commonly used in agriculture to improve crop yield and efficiency. Nanopesticides are one of a new strategies being used to address the problems of non-nanopesticides. Nanopesticides cover a wide variety of products, some of which are already on the market. They cannot be considered as a single entity; rather such nanoformulations combine several surfactants, polymers (organic), and metal nanoparticles (inorganic) in the nanometer size range. The lack of water solubility is one of the limiting factors in the development of crop-protecting agents. Microencapsulation has been used as a versatile tool for hydrophobic pesticides, enhancing their dispersion in aqueous media and allowing a controlled release of the active compound. Polymers often used in the nanoparticle production have been reported.

First of all, polycaprolactone and poly(lactic) acid nanospheres were used for encapsulation of the insecticide ethiprole. In this case, results indicated that nanospheres do not provide a controlled release of agrochemical active ingredients but, due to their small size, they enhanced the penetration in the plant compared to the classical suspension. In vivo experiments carried out with Egyptian cotton leafworm *Spodoptera littoralis*larvae indicated that the toxicity of nanoparticles of novaluron resembled that of the commercial formulation.

Nanomaterials serve equally as additives (mostly for controlled release) and active constituents. Controlled-release (CR) formulations of imidacloprid (1-(6 chloro-3-pyridinyl methyl)-N-nitro imidazolidin-2-ylideneamine), synthesized from polyethylene glycol and various aliphatic diacids using encapsulation techniques, have been used for efficient pest management in different crops. The bioefficacy of the prepared CR formulations and a commercial formulation were evaluated against major pests of soybean, namely stem fly, *Melanagromyza sojae* Zehntmer and white fly, *Bemisia tabaci* Gennadius. Most of the CR formulations of imidacloprid exhibited better control of the pests compared with its commercial formulations; however, of the CR formulations, poly(poly(oxyethylene-1000)-oxy suberoyl) amphiphilic polymer-based formulation performed better than others for controlling of both stem fly incidence and Yellow Mosaic Virus infestation transmitted by white fly. In addition, some of the developed CR formulations recorded higher yield over commercial formulation and control.

CR formulations of carbofuran and imidacloprid provided better or equal control against the aphid, *Aphis gossypii* and leafhopper, *Amrasca biguttula biguttula* Ishida on potato crop, than commercial formulations. The residue of carbofuran and imidacloprid in potato tuber and soils was not detectable at the time of harvesting in any one of the formulations. Nanomaterials including polymeric nanoparticles, iron oxide nanoparticles, gold nanoparticles, and silver ions have been exploited as pesticides. Researchers have reported various aspects of nanoparticle formulation, characterization, effect of their characteristics, and their applications in management of plant diseases.

Nanoparticles in insects and their potential for use in insect pest management have been reported. Nanotechnology in the management of polyphagous pest *Helicoverpa armigera* has been reported. The pediculocidal and larvicidal activity of synthesized silver nanoparticles using an aqueous leaf extract of*Tinospora cordifolia* showed maximum mortality against the head louse *Pediculus humanus* and fourth instarlarvae of *Anopheles subpictus* and *Culex-quinque fasciatus*. Synthesized silver nanoparticles possessed excellent antilice and mosquito larvicidal activity.

Nanoencapsulation helps slow release of a chemical to the particular host for insect pest control through release mechanisms that include dissolution, biodegradation, diffusion, and osmotic pressure with specific pH. Nanoparticles loaded with garlic essential oil proved effective against *Tribolium castaneum* Herbst. The use of amorphous nanosilica as biopesticide has been reported. Nanocopper particles suspended in water have been used since at least 1931, in a product known as Bouisol as fungicide in the growing of grapes and fruit trees.

In the research and development stage, nanosized agrochemicals or nanoagrochemicals are mostly nano-reformulations of existing pesticides and fungicides. Nanoformulations are generally expected to increase the apparent solubility of poorly soluble active ingredients, to release the active ingredient in a slow/targeted manner, and/or to protect against premature degradation. Nanopesticides offer a way to both control delivery of pesticide and achieve greater effects with lower chemical dose.

Agrochemical companies are reducing the particle size of existing chemical emulsions to the nanoscale, or are encapsulating active ingredients in nanocapsules designed to split open, for example, in response to sunlight, heat, or the alkaline conditions in an insect's stomach. The smaller size of nanoparticles and emulsions used in agrochemicals is intended to make them more potent. Many companies make formulations that contain nanoparticles within the 1 00–250 nm size range that are able to dissolve in water more effectively than existing ones, thus increasing their activity. Other companies employ suspensions of nanoscale particles (nanoemulsions), which can be either water-based or oil-based and contain uniform suspensions of pesticidal or herbicidal nanoparticles in the range of 200–400 nm.

Nanocapsules can enable effective penetration of herbicides through cuticles and tissues, allowing slow and constant release of the active substances. Viral capsids can be altered by mutagenesis to achieve different configurations and deliver specific nucleic acids, enzymes, or antimicrobial peptides acting against the parasites. Silver nanoparticles at 100 mg/kg inhibited mycelia growth and conidial germination on cucurbits and pumpkins against powdery mildew. Silver nanoparticles have received significant attention as a pesticide for agricultural applications. The potential of nanomaterials in insect pest management as modern approaches of nanotechnology, has been reported.

Treatment of *Bombyx mori* leaves with grasserie disease with ethanolic suspension of hydrophobic alumina–silicate nanoparticles significantly reduced the viral load. DNA-tagged gold nanoparticles are effective against *Spodoptera litura* and would therefore be a useful component of an integrated pest-management strategy. Development of nanobased viral diagnostics including kits can help to detect the exact strain of virus and identify differential proteins in healthy and diseased states during the infectious cycle and the stage of application of therapeutics to stop disease, thus increasing speed as well as power of disease detection.

Nanosilica has been successfully employed to control a range of agricultural insect/pest and ectoparasites in animals. Such nanoparticles get absorbed into cuticular lipids (used by insects to prevent death from desiccation) by physisorption and cause insect death by physical means when applied on leaves and stem surfaces. Antifungal activities of polymer-based copper nanocomposites against pathogenic fungi, and silica–silver nanoparticles against *Botrytis cinerea, Rhizoctonia solani, Calllectotrichum gloeosporioides*, *Bipolaris sorokiniana*, and *Magnaporthe grisea* have been reported. Copper nanoparticles in soda lime glass powder showed efficient antimicrobial activity against gram-positive and gram-negative bacteria and fungi. A novel photodegradable insecticide involving nanoparticles has been reported.

Specific nanoencapsulated pesticides will have the ability to kill targeted insects only, thereby reducing the effective dose when compared to traditional pesticides. Further, these are absorbed on the surface of the plant, facilitating a prolonged release that lasts for a longer time compared to conventional pesticides that wash away in the rain. Significant mortality of two insect pests, *Sarocladium oryzae* and *Rhyzopertha dominica*, after 3 days' exposure to nanostructured alumina-treated wheat was reported. Halloysite nanotube has potential to be applied as a nanocontainer for encapsulation of chemically and biologically active agents such as agromedicines and pesticides.

It is essential to remove weeds for increasing the yield of any crop and weeding using nanoherbicides is seen as an economically viable alternative. Conventional herbicides have proved highly effective in controlling weeds without damage to crops or environment. However, chemical weed management under rain-fed areas depends on the moisture availability during the application of herbicides. Lack of moisture limits the use and efficiency of the application. The nano-silicon carrier comprising diatom frustules (pore size 1–100 nm) has been used for delivery of pesticides and herbicides in plants as well as in hormonal waste-water treatment. CR formulation is superior to its counterpart and results in a higher yield and better crop quality. Such a formulation also finds use in active-agent herbicides, pesticides, and plant growth regulators.

The potential application of a layered single-metal hydroxide, particularly zinc-layered hydroxide, as the host for the preparation of a nanohybrid compound with a tunable CR property containing two herbicides simultaneously has been demonstrated. In this context, a nanohybrid containing both herbicides (4-(2,4-dichlorophenoxy) butyrate [DPBA] and 2-(3-chlorophenoxy) propionate [CPPA]) labeled as ZCDX was found a suitable host for the CR formulation of two herbicides, namely DPBA and CPPA, simultaneously.

The monophasic, well-ordered zinc-layered hydroxide nanohybrid containing two herbicides, CPPA and DPBA, was found to be composed of a higher loading of DPBA compared to CPPA between the zinc-layered hydroxide inorganic interlayers, with percentage contributions of 83.78 per cent and 16.22 per cent, respectively. The release rate of both CPPA and DPBA was found to be different, suggesting that the anionic guest molecules' sizes and the interactions between the host and guest could control the release kinetics.

Researchers reported a functional hybrid nanocomposite based on the intercalation of two herbicides' anions (2,4-dichlorophenoxy acetate and 4-chlorophenoxy acetate) with zinc–aluminum-layered double hydroxide. CR formulations of nanocomposites such as 4-chlorophenoxy acetate–zinc–aluminium-layered double hydroxide and 4-dichlorophenoxy acetate–zinc–aluminum-layered double hydroxide were reported. Researchers reported manganese carbonate core-shell nanoparticles loaded with pre-emergence herbicide pendimethalin programmed to release smartly based upon the requirements.

The field of nanotechnology opens up novel applications in agriculture. Nanoencapsulation is currently the most promising technology for protection of host plants against insect pests. With nanoencapsulation techniques it is possible to step down the chemical release under controlled situations, reducing the current application dosage and improving efficiency.

Nanoparticles can be used in the preparation of new formulations like pesticides, insecticides, and insect repellants. Researchers have reported nanosilver and titanium dioxide nanoparticle applications in management of plant diseases.

Fungicidal efficiency of sulfur nanoparticles against two phytopathogens, *Fusarium solani* (isolated from an infected tomato leaf, responsible for early blight and *Fusarium* wilt diseases) and *Venturia inaequalis*(responsible for the apple scab disease) has been reported.

Pheromones are naturally occurring volatile semiochemicals and are considered ecofriendly biological control agents. Pheromones immobilized in a nanogel exhibited high residual activity and excellent efficacy in an open orchard. Environment-friendly management of fruit flies involving pheromones for the reduction of undesirable pest populations responsible for decreasing yield and crop quality has been reported. Within this scope, the preparation of a nanogel from a pheromone, methyl eugenol using a low-molecular-mass gelator such as all-trans tri(p-phenylene vinylene) bis-aldoxime was reported.

The nanogel formation is possible due to the self-assembly of the gelator molecules in pheromone. The nanogel offers stability at open ambient conditions, reduced evaporation and sustained release of the pheromone, easy handling and transportation without refrigeration, and reduction in the frequency of pheromone recharging in the orchard. The protocol also depicts an easy sampling technique for the trapping of the pests in a guava orchard using the nanogels. Notably the involvement of the nanogelled pheromone brought about an effective management of *Bactrocera dorsalis*, a prevalent harmful pest for a number of fruits, including guava. Researchers studied the insecticidal activity of nanostructured alumina against two insect pests, *Sarocladium oryzae* (L.) and *Rhyzopertha dominica* (F.), which are major insect pests in stored food supplies throughout the world. Significant mortality was observed after 3 days of continuous exposure to nanostructured alumina-treated wheat. Therefore, compared to commercially available insecticides, inorganic nanostructured alumina may provide a cheap and reliable alternative for control of insect pests, and such studies may expand the frontiers for nanoparticle-based technologies in pest management.

Recent advances in nanoscale engineering have created a new class of particulate bionanotechnology that uses biomimicry to better integrate adjuvant and antigen. These pathogen-like particles can come from a variety of sources, ranging from fully synthetic platforms to biologically derived, self-assembling systems. By employing molecularly engineered targeting and stimulation of key immune cells, recent studies utilizing pathogen-like particles as vaccine delivery platforms have shown great promise against high-impact, unsolved vaccine targets ranging from bacterial and viral pathogens to cancer and addiction. Controlling pests and investigation of the plant pathology field on the basis of nanobiotechnology is challenging.

The beneficial effects of different nanosize metals on plant cells and seed germination has also been reported. These controversial findings require clarification in order to avoid confusion to the public. However, considering these positive and negative effects of nanoparticles, it is necessary to evaluate impact of nanoparticles on plant cells, especially seeds that grow in agricultural soils. It is possible to avoid culturing sensitive plants in these areas and replace them with plants that tolerate high concentrations of nanoparticles.

Nanofertilizers

Nanofertilizer technology is very innovative, and scant reported literature is available in the scientific journals. Substituting nanofertilizers for traditional methods of fertilizer application is a way to release nutrients into the soil gradually and in a controlled way, thus preventing autrification and pollution of water resources.

Treatment with TiO_2 nanoparticles on maize had a considerable effect on growth, whereas the effect of TiO_2 bulk treatment was negligible. Titanium nanoparticles increased light absorption and photo energy transmission. In another experiment, a compound of SiO_2 and TiO_2 nanoparticles increased the activity of nitrate reductase in soybeans and intensified plant absorption capacity, making its use of water and fertilizer more efficient. Iranian researchers have produced the nano-organic iron-chelated fertilizer that is environmentally sustainable. Nanofertilizers have unique features like ultrahigh absorption, increase in production, rise in photosynthesis, and significant expansion in the leaves' surface area.

The use of nanofertilizer leads to an increased efficiency of the elements, reduces the toxicity of the soil, to at least reach the negative effects caused by the consumption of excessive consumption of fertilizers, and reduces the frequency of application of fertilizers. The positive effect (uptake and the penetration of zinc oxide nanoparticles in the leaves) of zinc oxide nanoparticles on tomato plants opens an avenue for its potential use as a future nanofertilizer. Pot studies with foliar spray approach showed that plants sprayed with 20 mg mL"^{1} zinc oxide nanoparticle solution showed improved growth and biomass production as compared to control plants.

Fertilizers based on nanotechnology have the potential to surpass conventional fertilizers. Innovation in nanofertilizers requires a multidisciplinary approach. In nanofertilizers, nutrients can be encapsulated by nanomaterials, coated with a thin protective film, or delivered as emulsions or nanoparticles. Nanobased slow-release or CR fertilizers have the potential to increase the efficiency of nutrient uptake. Engineered nanoparticles are useful for mitigating the chronic problem of moisture retention in arid soils and enhancing crop production by increasing the availability of nutrients in the rhizosphere. Coating and binding of nano- and subnano-composites help to regulate the release of nutrients from the fertilizer capsule.

In this context, researchers showed that application of a nanocomposite consisting of nitrogen, phosphorus, potassium, micronutrients, mannose, and amino acids enhanced the uptake and use of nutrients by grain crops. Zinc–aluminium layered double-hydroxide nanocomposites have been employed for the controlled release of chemical compounds that act as plant growth regulators. Urea-modified hydroxyapatite nanoparticle-encapsulated *Gliricidia sepium* nanocomposite displayed a slow and sustained release of nitrogen over time at three different pH values. Nanoporous zeolite based on nitrogen fertilizer can be used as alternate strategy to improve the efficiency of nitrogen use in crop production systems. As superfertilizer, carbon nanotubes were found to penetrate tomato seeds and affect their germination and growth rates. Analytical methods indicated that the carbon nanotubes penetrated the thick seed coat and supported water uptake inside seeds. Encapsulation of fertilizers within a nanoparticle is done in three ways: the nutrient can be 1) encapsulated inside nanoporous materials, 2) coated with thin polymer film, or 3) delivered as particles or emulsions of nanoscale dimensions. The roles of nanofertilizers in sustainable agriculture have been reported.

NANOBIOTECHNOLOGY IN AGRI-FOOD PRODUCTION

Nanobiotechnology opportunities include food, agriculture and energy applications. In the food processing industries, a few of the most common usages of nanobiotechnology in quality monitoring of food products may be enumerated as nanosensors/nanobiosensors and bacteria identification. The nanosensors can be utilized to detect the presence of insects or fungus accurately inside the stored grain bulk in storage rooms. Researchers suggested models for use of nanobiotechnology, either on a standalone basis or through complementarity with the existing technologies. In 2004, the researchers had been able to alter rice colour from purple to green.

Cellular "injection" with carbon nanofibres containing foreign DNA has been used to genetically modify golden rice. Nanobiotechnology provided industry with new tools to modify genes and even produce new organisms. This is due to the fact that it enables nanoparticles, nanofibres, and nanocapsules to carry foreign DNA and chemicals that modify genes. In addition, novel plant varieties may be developed using synthetic biology (a new branch that draws on the techniques of genetic engineering, nanotechnology, and informatics). In a recent breakthrough in this area, researchers completely replaced the genetic material of one bacterium with that from another – transforming it from one species to another. Nanotechnology possesses the potential to augment agricultural productivity through genetic improvement of plants and animals along with cellular level delivery of genes and drug molecules to specific sites in plants and animals.

Using a medicinally rich vegetable crop, bitter melon, researchers demonstrated the accumulation of carbon-based nanoparticle Fullerol ($C_{60}(OH)_{20}$) in tissues and cells of root, stem, petiole, leaf, flower and fruit at particular concentrations, as the causal factor of increase in biomass yield, fruit yield, and phytomedicine content in fruits. Fullerol treatment resulted in increases of up to 54 per cent in biomass yield and 24 per cent in water content. Increases of up to 20 per cent in fruit length, 59 per cent in fruit number, and 70 per cent in fruit weight led to an improvement of up to 128 per cent in fruit yield.

Further, contents of two anticancer phytomedicines, cucurbitacin-B and lycopene, were enhanced up to 74 per cent and 82 per cent, respectively, and contents of two antidiabetic phytomedicines, charantin and insulin, were augmented up to 2 0 per cent and 9 1 per cent, respectively. Chemists have successfully made DNA crystals by producing synthetic DNA sequences that can self-assemble into a series of three-dimensional triangle-like patterns. When multiple helices are attached through single-stranded sticky ends, a three-dimensional crystal is formed. This technique helps in improving important crops by organizing and linking carbohydrates, lipids, proteins, and nucleic acids to this crystal.

Chemically coated mesoporous silica nanoparticles help in delivering DNA and chemicals into isolated plant cells. The coating triggers the plant to take the particles through the cell walls, where the genes are inserted and activated in a precise and controlled manner, without any toxic side or after effects. This technique has been applied to introduce DNA successfully to plants, including tobacco and corn plants.

Nanosensors/nanobiosensors in agri-food production

Nanobiosensors can be effectively used for sensing a wide variety of fertilizers, herbicide, pesticide, insecticide, pathogens, moisture, soil pH, and their controlled use can support sustainable agriculture for enhancing crop productivity. Precision farming, with the help of smart sensors, could increase productivity in agriculture, as this technology provides farmers with better fertilization management, reduction of inputs, and better management of time and the environment.

Nanosensors and nanobased smart delivery systems could help in the efficient use of agricultural natural resources like water, nutrients, and chemicals through precision farming. Precision farming's enabling technologies include satellite-positioning systems, geographic information systems, and remote sensing devices that could remotely detect crop pests or evidence of stress such as drought. Nanosensors dispersed in the field can also detect the presence of plant viruses and other crop pathogens, and the level of soil nutrients. Levels of environmental pollution can be evaluated quickly by nano-smart dust (the use of tiny wireless sensors and transponders) and gas sensors. Nanobarcodes and nano-processing could also be used to monitor the quality of agricultural produce. Nanotechnology-based plant regulation of hormones such as auxin helps scientists understand how plant roots adapt to their environment, especially to marginal soils.

The development of sensors/biosensors based on specific interactions makes atomic force spectroscopy more effective in detecting enzyme-inhibiting herbicides. A nanobiosensor based on an atomic force microscopy tip functionalized with the acetolactate synthase enzyme was successfully detected for the herbicide metsulfuron-methyl (an acetolactate synthase inhibitor) through the acquisition of force curves. Bionanosensors also allow the more quantification and rapid detection of bacteria and viruses, thereby increasing the safety of the food for the customer.

Noble metal (palladium, platinum, and gold)/DNA/single-walled carbon nanotube (SWCNT) hybrid nanostructure-based gas sensor arrays were fabricated by means of inkjet printing of metal ion-chelated DNA/SWCNTs on microfabricated electrodes, followed by electroless deposition to reduce metal ions to metal. DNA served as a dispersing agent to effectively solubilize pristine SWCNTs in water and as metal ion-chelating centers for the formation of nanoparticles. The results on the sensitivity and selectivity of the gas sensors towards various gases such as H_2, H_2S, NH_3, and NO_2 indicated the enhancement of the sensitivity and selectivity towards certain analytes by functionalizing with different metal nanoparticles (eg, Pd/DNA/SWCNTs for H_2 and H_2S). The combined responses give a unique pattern or signature for each analyte by which the system can identify and quantify an individual gas.

Nanosensors are expected to impact agricultural, food, and environmental sectors. The Nanotechnology Signature Initiative "Nanotechnology for Sensors and Sensors for Nanotechnology: Improving and Protecting Health, Safety, and the Environment" is the fifth to be launched by agencies of the National Nanotechnology Initiative. Portable nanodevices can rapidly detect insects, diseases, pathogens, chemicals, and contaminants and can result in faster treatments.

Nanosensors based on using electrochemically functionalized SWCNTs with either metal nanoparticles or metal oxide nanoparticles, and metal oxide nanowires and nanotubes for gases such as ammonia, nitrogen oxides, hydrogen sulfide, sulfur dioxide, and volatile organics have potential application in monitoring agricultural pollutants for the assessment of impacts of these pollutants on biological and ecological health and in increase of crop productivity and reducing land burden. Researchers addressed the fabrication, functionalization, assembly/alignment, and sensing applications of field-effect transistors based on carbon nanotubes, silicon nanowires, and conducting polymer nanowires. Further, they evaluated how such sensors have been used for detection of various biological molecules and how such devices have enabled the achievement of high sensitivity and selectivity with low detection limits.

Nanotechnology-enabled devices will increase the use of sensors linked to global positioning systems for real-time monitoring of crops. In the field of sensor research and development, bionanotechnology is poised to make significant contributions and has the potential to radically alter the way sensors are designed, constructed, and implemented. Biomimetic nanosensor designs based on immobilized tyrosinase for determination of toxic compounds and smart biosensors for determination of mycotoxines were reported. Biosensor design showed good compatibility between membranes and enzymes without a change of the conformation of the enzyme molecule, and binding always takes place outside the enzyme active centers.

Carbon–ceramic electrode modified with multi-walled carbon nanotubes–ionic liquid nanocomposite was used for electrochemical determination of the food dyes, sunset yellow and tartrazine, in food and beverage samples.

Bibliography

A.S. Kadale, M.S. Pendke, S.D. Payal and A.A. Waghmare: *Integrated Watershed Approach : For Sustainable Crop Production*, Agrotech Publication, Delhi, 2011.

Ajoy Gangopadhyay: *Assessment of Improved Crop Production Technologies*, Gene Tech Books, Delhi, 2007.

Ajoy Gangopadhyay: *Crop Production Systems Management*, Gene-Tech Books, Delhi, 2007.

Beena Nair; Krishna Pal Singh and Prem Chand: *Fundamentals of Vegetable Crop Production*, Scientific Publishers, Delhi, 2014.

Benjamin Wolf: *Diagnostic Techniques for Improving Crop Production*, CBS Publication, Delhi, 2003.

C.P. Malik: *Genetic Engineering : A New Hope for Crop Production and Improvement*, Aavishkar Publication, Delhi, 2010.

Charlotte Jhonson: *Crop Production in Stress Environment*, Oxford Book Company, Delhi, 2010.

D P Singh; V S Tomar; R K Behi; S D Upadhyaya; M S Bhale and D Khare: *Crop Production in Stress Environments: Genetic and Management Options*, Agrobios Publication, Delhi, 2007.

E.J. Russell: *Manuring for Higher Crop Production*, Agrobios Publication, Delhi, 2004.

Fred C. Blank: *Essential Aspects of Agricultural Crop Production*, Allied Publication, Delhi, 1998.

G L Maliwal: *Crop Production with Waste Water*, Agrotech Publication, Delhi, 2004.

G.S. Chaturvedi, P.C. Ram and A.K. Singh: *Physiological Basis of Crop Production and Improvement*, Aavishkar Publication, Delhi, 2009.

L L Somani: *Crop Production With Saline Water*, Agrobios Publication, Delhi, 2007.

L.L. Somani and S.C. Bhandari: *Organic Recycling and Bioinoculants : For Sustainable Crop Production*, Agrotech Publication, Delhi, 2007.

Lucius L. Van Slyke: *Fertilizers and Crop Production*, Agrobios Publication, Delhi, 2010.

P C Bandyopadhyay: *Breeding and Crop Production*, Gene Tech Books, Delhi, 2007.

P.C. Ram and G.S. Chaturvedi: *Crop Production Under Diverse Environments*, Pointer Publication, Delhi, 2009.

Panda. S. C.: *Modern Concepts and Advances Principles in Crop Production*, Agrobios Publication, Delhi, 2012.

Parameshwar Singh: *Encyclopaedia of Scientific Crop Production*, Anmol Publication, Delhi, 2010.

R K Mishra: *Livestock-Crop Production Systems and Livelihood Development*, Atlantic Publication, Delhi, 2007.

Ranjan Das: *Climate Variability and its Impact on Crop Production*, Agrotech Publishing Academy, Delhi, 2013.

S C Panda: *Crop Production and Tillage*, Agrobios Publication, Delhi, 2007.

S K Gupta: *Horticultural Crop Production*, Rajat Publication, Delhi, 2008.

S.K. Gupta: *Crop Production in Waterlogged Saline Soils*, Scientific Publication, Jaipur, 1997.

Sampat Nehra: *Arbuscular Mycorrhizae in Crop Production : Professor P.C. Trivedi Festschrift Volume*, Pointer Publishers, Delhi, 2013.

Shagufta: *Crop Production and Tillage*, A.P.H. Publication, Delhi, 2011.

Sham S. Goyal, Surinder K. Sharma and D. William Rains: *Crop Production in Saline Environments : Global and Integrative Perspectives*, International Book, Delhi, 2004.

Shamsher S. Narwal: *Allelopathy in Crop Production*, Scientific Publication, Jaipur, 2004.

Supriya Kumar Bhattacharjee: *Ornamental Crop Production Technology*, Pointer Publication, Delhi, 2006.

Vikas Ahlluwalia: *Environmental and Ecological Aspects of Crop Production*, Paragon International Publication, Delhi, 2007.

Index

G

H

I

L

M

N

O

P

R

S

T

V

W